Anne-Katharina Jappsen

Present-day and Early Star Formation

Anne-Katharina Jappsen

Present-day and Early Star Formation

A Study on Rotational and Thermal Properties of Star-Forming Molecular Clouds

Südwestdeutscher Verlag für Hochschulschriften

Impressum/Imprint (nur für Deutschland/ only for Germany)
Bibliografische Information der Deutschen Nationalbibliothek: Die Deutsche Nationalbibliothek verzeichnet diese Publikation in der Deutschen Nationalbibliografie; detaillierte bibliografische Daten sind im Internet über http://dnb.d-nb.de abrufbar.
Alle in diesem Buch genannten Marken und Produktnamen unterliegen warenzeichen-, marken- oder patentrechtlichem Schutz bzw. sind Warenzeichen oder eingetragene Warenzeichen der jeweiligen Inhaber. Die Wiedergabe von Marken, Produktnamen, Gebrauchsnamen, Handelsnamen, Warenbezeichnungen u.s.w. in diesem Werk berechtigt auch ohne besondere Kennzeichnung nicht zu der Annahme, dass solche Namen im Sinne der Warenzeichen- und Markenschutzgesetzgebung als frei zu betrachten wären und daher von jedermann benutzt werden dürften.

Verlag: Südwestdeutscher Verlag für Hochschulschriften Aktiengesellschaft & Co. KG
Dudweiler Landstr. 99, 66123 Saarbrücken, Deutschland
Telefon +49 681 37 20 271-1, Telefax +49 681 37 20 271-0, Email: info@svh-verlag.de
Zugl.: Potsdam, Germany, Potsdam University, Diss., 2006

Herstellung in Deutschland:
Schaltungsdienst Lange o.H.G., Berlin
Books on Demand GmbH, Norderstedt
Reha GmbH, Saarbrücken
Amazon Distribution GmbH, Leipzig
ISBN: 978-3-8381-0417-1

Imprint (only for USA, GB)
Bibliographic information published by the Deutsche Nationalbibliothek: The Deutsche Nationalbibliothek lists this publication in the Deutsche Nationalbibliografie; detailed bibliographic data are available in the Internet at http://dnb.d-nb.de.
Any brand names and product names mentioned in this book are subject to trademark, brand or patent protection and are trademarks or registered trademarks of their respective holders. The use of brand names, product names, common names, trade names, product descriptions etc. even without a particular marking in this works is in no way to be construed to mean that such names may be regarded as unrestricted in respect of trademark and brand protection legislation and could thus be used by anyone.

Publisher:
Südwestdeutscher Verlag für Hochschulschriften Aktiengesellschaft & Co. KG
Dudweiler Landstr. 99, 66123 Saarbrücken, Germany
Phone +49 681 37 20 271-1, Fax +49 681 37 20 271-0, Email: info@svh-verlag.de

Copyright © 2009 by the author and Südwestdeutscher Verlag für Hochschulschriften Aktiengesellschaft & Co. KG and licensors
All rights reserved. Saarbrücken 2009

Printed in the U.S.A.
Printed in the U.K. by (see last page)
ISBN: 978-3-8381-0417-1

Contents

Abstract		7
1 Introduction		9
2 Observations of Star-forming Regions		11
2.1	Physical Properties of Molecular Clouds	13
2.2	Thermal Properties of Star-forming Clouds	15
2.3	The IMF	17
2.4	The Angular Momentum Problem	21
3 Modeling Star Formation in Different Epochs		23
3.1	Cosmological Preliminaries	23
	3.1.1 The Standard Model	23
	3.1.2 Hierarchical Structure Formation	25
3.2	Star Formation Theory - Two Short Histories	28
	3.2.1 Early Star Formation	28
	3.2.2 Current Star Formation	29
3.3	Gravoturbulent Fragmentation	29
	3.3.1 Self-Gravitating Hydrodynamics	30
	3.3.2 Jeans Criterion	31
	3.3.3 Turbulence	31
	3.3.4 Interstellar Turbulence	33
3.4	Equation of State	35
	3.4.1 Isothermal Equation of State	35

		3.4.2	Polytropic Equation of State	35
	3.5	\multicolumn{2}{l}{Timescales . . . .}	38

4 The Numerical Scheme — 40
   4.1 Smoothed Particle Hydrodynamics 40
   4.2 Resolution Issues 42
   4.3 Sink Particles 43
       4.3.1 The Concept 43
       4.3.2 Sink Particles in a Parallel Code 44
   4.4 Turbulent Driving 49
   4.5 Chemistry and Cooling 50
   4.6 Periodic Boundary Conditions 54

5 Angular Momentum Evolution — 56
   5.1 Initial Conditions and Model Parameters 57
       5.1.1 Physical scaling and naming convention 58
   5.2 Molecular Cloud Clumps and Prestellar Cores 59
   5.3 Protostars and Protostellar Systems 60
       5.3.1 Example of the Angular Momentum Evolution of a Protostellar Core in a Cluster 61
       5.3.2 Statistical Correlation between Specific Angular Momentum and Mass — 62
       5.3.3 Dependence of the Specific Angular Momentum on the Environment . 64
   5.4 Loss of Angular Momentum during Collapse 65
   5.5 Orientation of Angular Momentum Vector 65
   5.6 Summary and Conclusions 66

6 Non-isothermal Gravoturbulent Fragmentation — 77
   6.1 Model Parameters 78
   6.2 Gravoturbulent Fragmentation in Polytropic Gas 79
   6.3 Dependence of the Characteristic Mass on the Equation of State 85
   6.4 Dependence of the Characteristic Mass on Environmental Parameters 87
       6.4.1 Dependence on Realization of the Turbulent Velocity Field 87

		6.4.2 Dependence on the Scale of Turbulent Driving	89
	6.5	Summary	89

7 Cooling and Collapse of Ionized Gas in Small Protogalactic Halos — 92
- 7.1 Initial Conditions 93
- 7.2 Zero Metallicity Gas 97
  - 7.2.1 Dependence of Cooling and Collapse on Halo Mass 97
  - 7.2.2 Dependence of Cooling and Collapse on UV Background 99
- 7.3 Low Metallicity Gas 102
  - 7.3.1 Dependence of Cooling and Collapse on Metallicity 104
  - 7.3.2 Dependence of Cooling and Collapse on Metallicity and UV Background 106
- 7.4 Summary and Discussion 109

8 Summary and Future Prospects — 114

A Physical Units and Constants — 118

B Primordial and Low Metallicity Gas Chemistry — 119

Index — 143

List of Figures

2.1	Millimeter continuum mosaic of the ρ-Ophiuchus cloud	12
2.2	Comparison of the Class 1 source EL 29 in ρ-Oph and the Class 0 source L 1527 in Taurus	13
2.3	The distribution of clump line widths for the clumps within the M17 SW cloud core.	14
2.4	Theoretical and observed temperatures of interstellar clouds as a function of density	16
2.5	Measured stellar mass function	18
3.1	Local Jeans mass as a function of density	37
4.1	Column density and velocity field for runs with different realizations of the Gaussian velocity field	46
4.2	Column density and isodensity contours for runs with different realizations of the Gaussian velocity field	47
4.3	Column density and isodensity contours for runs with different realizations of the Gaussian velocity field (self-gravity added)	48
4.4	Root mean square velocity of the gas particles as a function of time	50
5.1	Distribution of specific angular momenta of prestellar cores	69
5.2	Distribution of specific angular momenta of the protostars or protostellar systems	70
5.3	Absolute value of specific angular momentum as (a) a function of mass and as (b) a function of time for five different protostellar objects with approximately equal final masses	71
5.4	Absolute values of specific angular momenta as a function of mass	72
5.5	Distribution of β of the collapsed protostellar cores	73
5.6	Average specific angular momentum of protostellar objects in different turbulent environments as a function of the associated Mach number	74

5.7 The correlation of specific angular momenta of different protostellar cores with respect to their orientations as a function of distance between them. 75

5.8 Orientation of the angular momenta and spatial distribution of the protostellar cores 76

6.1 Temperature as a function of density for four runs with different critical densities n_c. 81

6.2 Column density distribution of the gas and location of identified protostellar objects 82

6.3 Temporal evolution of the number of protostellar objects and of the ratio of accreted gas mass to total gas mass for models R5..8k2b. 84

6.4 Mass spectra of protostellar objects for models R5..6k2b, model R7k2L and model R8k2L 86

6.5 Median mass of the protostellar objects over critical density for models R5..6k2b, model R7k2L and model R8k2L 87

6.6 Median mass of protostellar objects over critical density at different evolutionary phases 88

7.1 Projection of the hydrogen nuclei number density in x-y plane. 98

7.2 The time evolution of the number density of hydrogen nuclei within the scale radius r_s of the dark matter halo for three suites of runs in different ranges of the dark matter halo mass 100

7.3 The time evolution of the gas number density within the scale radius r_s of the dark matter potential. 101

7.4 Same as Figure 7.3, but for the central temperature of the gas. 102

7.5 The radial dependence of the number density of the gas for models ZM25m and LM25m at several times. 103

7.6 Time evolution of the radial dependence of various parameters for models ZM25m and LM25m at several times. 105

7.7 The radial dependence of various metal abundances for model LM25m at several times. 107

7.8 Time evolution of the central gas density comparing runs with zero metallicity and runs with low metallicity 108

7.9 Same as Figure 7.8, but for the central temperature of the gas. From Jappsen et al. (2007). 109

7.10 Temperature and density at which the cooling time due to fine structure emission, $t_{\rm cool,fs}$, equals the free-fall time 110

List of Tables

2.1	Physical properties of interstellar clouds	15
2.2	Characteristic values of specific angular momentum	21
5.1	Parameters of the simulations concerning the angular momentum evolution	57
6.1	Parameters of the non-isothermal simulations	80
7.1	Parameters of simulations of gas in protogalactic dark matter halos	93
7.1	–Continued	94
7.1	–Continued	95
7.2	Physical state of the densest gas within the scale radius r_s after 1 Hubble time.	104
A.1	Physical Units	118
A.2	Physical Constants	118
B.1	A list of all the gas-phase reactions included in our chemical model.	120
B.1	–Continued	122
B.1	–Continued	123
B.2	A list of all the grain surface reactions included in our chemical model.	123

Abstract

We investigate the rotational and thermal properties of star-forming molecular clouds using hydrodynamic simulations. Stars form from molecular cloud cores by gravoturbulent fragmentation. Understanding the angular momentum and the thermal evolution of cloud cores thus plays a fundamental role in completing the theoretical picture of star formation. This is true not only for current star formation as observed in regions like the Orion nebula or the ρ-Ophiuchi molecular cloud but also for the formation of stars of the first or second generation in the universe.

In this thesis we show how the angular momentum of prestellar and protostellar cores evolves and compare our results with observed quantities. The specific angular momentum of prestellar cores in our models agree remarkably well with observations of cloud cores. Some prestellar cores go into collapse to build up stars and stellar systems. The resulting protostellar objects have specific angular momenta that fall into the range of observed binaries. We find that collapse induced by gravoturbulent fragmentation is accompanied by a substantial loss of specific angular momentum. This eases the "angular momentum problem" in star formation even in the absence of magnetic fields.

The distribution of stellar masses at birth (the initial mass function, IMF) is another aspect that any theory of star formation must explain. We focus on the influence of the thermodynamic properties of star-forming gas and address this issue by studying the effects of a piecewise polytropic equation of state on the formation of stellar clusters. We increase the polytropic exponent γ from a value below unity to a value above unity at a certain critical density. The change of the thermodynamic state at the critical density selects a characteristic mass scale for fragmentation, which we relate to the peak of the IMF observed in the solar neighborhood. Our investigation generally supports the idea that the distribution of stellar masses depends mainly on the thermodynamic state of the gas.

A common assumption is that the chemical evolution of the star-forming gas can be decoupled from its dynamical evolution, with the former never affecting the latter. Although justified in some circumstances, this assumption is not true in every case. In particular, in low-metallicity gas the timescales for reaching the chemical equilibrium are comparable or larger than the dynamical timescales.

In this thesis we take a first approach to combine a chemical network with a hydrodynamical code in order to study the influence of low levels of metal enrichment on the cooling and collapse of ionized gas in small protogalactic halos. Our initial conditions represent protogalaxies forming within a fossil H ii region – a previously ionized H ii region which has not yet had time to cool and recombine. We show that in these regions, H_2 is the dominant and most effective coolant, and that it is the amount of H_2 formed that controls whether or not the gas can collapse and form stars. For metallicities $Z \leq 10^{-3}\, Z_\odot$, metal line cooling alters the density and temperature evolution of the gas by less than 1% compared to the metal-free case at densities below $1\,\mathrm{cm}^{-3}$ and temperatures above $2000\,\mathrm{K}$. We also find that an external ultraviolet background delays or suppresses the cooling and collapse of the gas regardless of whether it is metal-enriched or not. Finally, we study the dependence of this process on redshift and mass of the dark matter halo.

Chapter 1

Introduction

Stars are a fundamental part of the cosmic circuit of matter. They are also a primary source of astronomical information and hence are essential for our understanding of the universe and the physical processes that govern its evolution. Stars emit radiation that provides us with information about their outer layers as well as their interiors. The life time of a star is determined by its mass and lies between a million and 35 billion years. Stars in their main sequence phase are hot massive dense gas spheres emitting radiation produced in their centers from nuclear fusion processes. As a result of the nuclear reactions the star produces metals (elements heavier than H, He). This process is the main energy source of the star and governs its main sequence phase. A star reaches this stage, following a period of gravitational contraction, as soon as the conditions in its center become hot and dense enough to start burning hydrogen. The gravitational contraction phase of a young star is qualitatively understood and there are theoretical models predicting its evolution towards the main sequence (pre-main-sequence tracks). In the final phase of its lifetime the star looses large amounts of material due to stellar winds and the supernova explosion. With gas and dust lost by the dying star new objects can form. The heaviest elements are produced during the passage of the final supernova shockwave through the outer layers of the most massive stars. Depending on the main-sequence mass of the star, its remanent is either a white dwarf, a neutron star or a black hole. To reach the present-day chemical abundances observed in our solar system, material had to go through many cycles of stellar birth and death.

For roughly the last century we have known that clouds of gas and dust are the sites where stars form. Advances in radio and infrared astronomy have made it possible to gain more knowledge about the interior of star-forming clouds. Nevertheless the very process of assembling gas to form stars still poses questions that star formation theory tries to answer. In Chapter 2 we therefore outline the observed properties of star-forming interstellar clouds as well as the distribution of stellar masses that form in these regions. The theoretical background is given in Chapter 3 where we introduce the concept of gravoturbulent fragmentation and give a short overview of the history of star formation theory. In Chapter 4 we describe the numerical method used and present our additions and improvements to the hydrodynamical code. Using the results of our simulations we present in Chapter 5 work that follows the angular momentum evolution of the objects formed and we compare our results with suitable observations. In Chapter 6 we discuss the influence of a non-isothermal equation of state on the fragmentation behavior of the star-forming gas and on the distribution of masses in the stellar clusters formed. The chemical and cooling properties of the gas is not only important in regions of recent star formation but also in the formation of stars of the first or second generation which have formed from low metallicity gas. In Chapter 7 we describe our simulations of ionized gas in small protogalactic halos. We examine specifically the cooling properties and the collapse properties of the gas in dependence on metallicity, the presence of an ultraviolet background and other parameters. In Chapter 8 we summarize and give an outlook on work that we will do in the future to strengthen the theory of present and early star formation.

Chapter 2

Observations of Star-forming Regions

All present-day star formation takes place in molecular clouds (see, for example, Blitz, 1993; Williams et al., 2000), so it is vital to understand the properties, dynamical evolution, and fragmentation behavior of molecular clouds in order to understand star formation. Molecular clouds are density enhancements in interstellar matter dominated by molecular H_2, rather than the atomic or the ionized H, typical of the rest of the interstellar medium (ISM). The formation of molecular hydrogen is due to the dust opacity of molecular clouds and due to the self-shielding of H_2 from UV radiation that elsewhere dissociates the molecules. In the plane of the Milky Way, interstellar gas has been extensively reprocessed by stars, so the metallicity is close to the solar value $Z_\odot$, while in other galaxies with lower star formation rates, the metallicity can be as little as $10^{-3} Z_\odot$. The refractory elements condense into dust grains, while others form molecules. The properties of the dust grains change as the temperature drops within the cloud, probably due to the freezing out of volatiles such as water and ammonia onto dust grains (Goodman et al., 1995). This has important consequences for the radiation transport properties and the optical depth of the clouds (Tielens, 1991; van Dishoek et al., 1993). The presence of heavier elements such as carbon, nitrogen and oxygen determines the heating and cooling processes in molecular clouds by fine structure lines (Genzel, 1991).

In addition, continuum emission from dust as well as emission and absorption lines caused by molecules formed from these elements are the observational tracers of cloud structure, as cold molecular hydrogen is very difficult to observe. Observations with radio and submillimeter telescopes mostly concentrate on the thermal continuum from dust and the rotational transition lines of carbon, oxygen, and nitrogen molecules (e.g. CO, NH_3, or H_2O). By now, several hundred different molecules have been identified in the interstellar gas. An overview of the application of different molecules as tracers for different physical conditions can be found in the reviews by van Dishoek et al. (1993), Langer et al. (2000), and van Dishoek & Hogerheijde (2000).

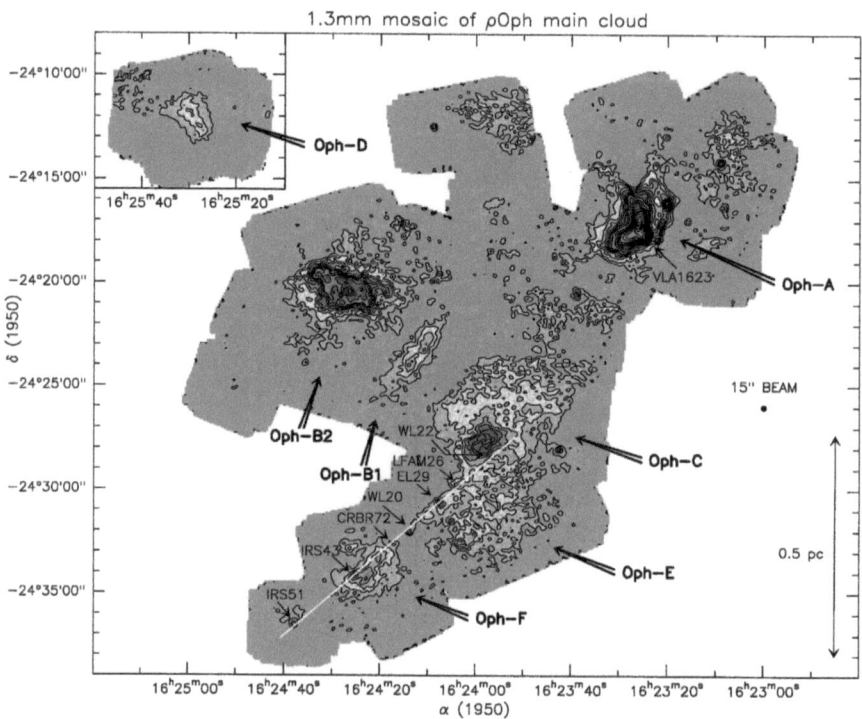

Figure 2.1: Millimeter continuum mosaic of the ρ-Ophiuchus cloud with the dense cores Oph-A, Oph-B1, Oph-B2, Oph-C, Oph-D, Oph-E, Oph-F. From Motte et al. (1998).

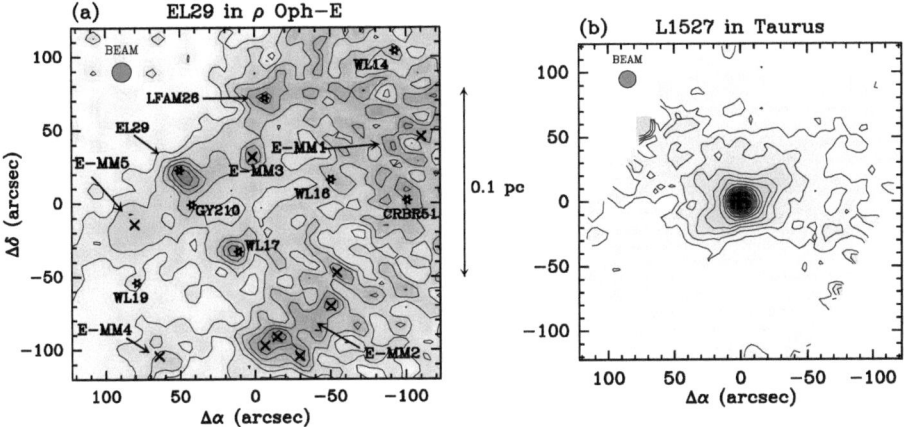

Figure 2.2: Comparison of the immediate 1.3 mm continuum environments of the Class 1 source EL 29 in ρ-Oph and the Class 0 source L 1527 in Taurus. From Motte et al. (1998).

2.1 Physical Properties of Molecular Clouds

The Interstellar Medium (ISM) consists of gas in all states (atoms, molecules, ions) and dust grains. The dust component accounts for only $1-2\%$ of the mass of the interstellar medium. The gas consists mainly of hydrogen and helium. A small percentage of the gas mass is in the form of heavier elements. Emission-line maps reveal clumps and filaments on all scales accessible by present-day telescopes. Typical parameters of different regions in molecular clouds are listed in Table 2.1, adapted from Cernicharo (1991). The largest molecular structures considered to be single objects are giant molecular clouds (GMCs), which have masses of $10^5 - 10^6\,M_\odot$ and extend over few tens of parsecs. The smallest observed structures are protostellar cores, with masses up to a few solar masses and sizes of $\leq 0.1\,\mathrm{pc}$, as well as less dense clumps of similar size. Star formation always occurs in the densest regions within a cloud, so only a small fraction of molecular cloud matter is actually involved in building up stars, while the bulk of the material remains at lower densities.

The hierarchy of clumps and filaments spans all observable scales extending down to individual protostars studied with millimeter-wavelength interferometry. At each level the molecular cloud appears clumpy and highly structured. When observed with higher resolution, each clump breaks up into a filamentary network of smaller clumps. Unresolved features exist even at the highest resolution. In Figure 2.1 we show ρ-Ophiuchi as an example of a star-forming region and its structure. Figure 2.2a also shows that density enhancements in this region are usually connected with clumps and young stellar objects denoted by crosses and star markers, respectively. The environment in ρ-Ophiuchi is highly structured, and that in the Taurus molecular cloud is much more quiescent. Figure 2.2b shows a prestellar object in this region.

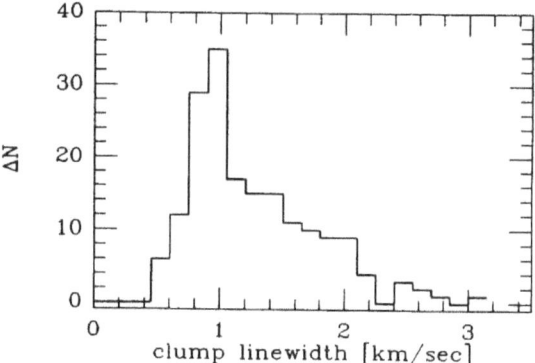

Figure 2.3: The distribution of clump line widths for 179 individual clumps within the M17 SW cloud core. From Stutzki & Guesten (1990).

The highly supersonic line-widths that are observed in molecular clouds are always wider than implied by the excitation temperature of the molecules. This is interpreted as the result of bulk motion associated with turbulence (see Section 3.3.3 and see discussion in Falgarone & Phillips, 1990). Figure 2.3 shows the distribution of clump line-widths for the M17 SW clumps within the cloud core. In this example the derived velocity dispersion is higher than the thermal line width of $C^{18}O$, $\Delta v_{\text{therm}} = 0.27\,\text{km}\,\text{s}^{-1} \times (T/50\,\text{K})^{1/2}$, but close to or above the line width expected for sonic turbulence, $\Delta v_{\text{sonic}} = 1.3\,\text{km}\,\text{s}^{-1} \times (T/50\,\text{K})^{1/2}$ (Stutzki & Guesten, 1990). Expanding H ii regions and supernovae are mechanisms for exciting turbulent motions in GMCs (Mac Low & Klessen, 2004). Some of the turbulent motions are generated by stars within these clouds, through their winds and outflows. The winds, outflows and radiation from the most massive stars probably limit the lifetime of molecular clouds by unbinding and dispersing the gas. It is likely that in their relatively short life times ($< 10^7$ yr) molecular clouds never reach a state of dynamical equilibrium (Ballesteros-Paredes et al., 1999a; Elmegreen, 2000b), but are rather transient compressed features of the turbulent flow of the interstellar medium.

An alternate description of the ISM is based on fractals. High spatial dynamic range observations of molecular clouds show exceedingly complex patterns that appear to defy a simple description in terms of clouds, clumps and cores (Falgarone et al., 1998; Lazarian & Pogosyan, 2000). Molecular clouds show self-similar structures down to the scales of prestellar objects (Chappell & Scalo, 2001).

There is also evidence for magnetic fields threading the giant molecular clouds. Polarimetry and Zeeman splitting measurements give average magnetic field strength of $10\,\mu\text{G}$ but measurements are difficult (Verschuur, 1995; Troland et al., 1996; Crutcher, 1999; Wada et al., 2000; Crutcher et al., 2003). It is under current discussion whether these relatively weak fields can stabilize molecular clouds as a whole or not (e.g. McKee et al., 1993; Shu et al., 1999; Mac Low & Klessen, 2004). Molecular Clouds are the birthplaces of all known young stars. They provide the initial conditions for star formation. Myers (1999) describes the stellar and prestellar content of molecular clouds as follows: "They host Young Stellar Objects (YSOs) in a wide range of evolutionary states; from Class 0 protostars some 10^{-2} Myr old, deriving most of their luminosity from gravitational infall, to T-Tauri stars a few Myr old, deriving their luminosity from quasi-static contractions. They also host stars in a wide range of spatial groupings; from isolated single stars as in Taurus (see Figure 2.2b), having no known neighbors within a few pc, to rich star clusters as in Orion, having a few thousand stars within a few pc. The masses of the stars in GMCs range from $0.1 - 30\,\mathrm{M}_\odot$, nearly the whole range of known stellar masses."

Table 2.1: Physical properties of interstellar clouds

Cloud Type	Size [pc]	Density $n(H_2)$ [cm^{-3}]	Mass [$M_\odot$]	Temperature [K]	Linewidth [km s^{-1}]
Giant Molecular Cloud Complex	$10 - 60$	$100 - 500$	$10^4 - 10^6$	$7 - 15$	$5 - 15$
Examples: W51, W3, M17, Orion-Monoceros, Taurus-Auriga-Perseus Complex					
Molecular Cloud	$2 - 20$	$10^2 - 10^4$	$10^2 - 10^4$	$10 - 30$	$1 - 10$
Examples: L1641, L1630, W33, W3A, B227, L1495, L1529					
Star-Forming Clump	$0.1 - 2$	$10^3 - 10^5$	$10 - 10^3$	$10 - 30$	$0.3 - 3$
Protostellar Core	≤ 0.1	$> 10^5$	$0.1 - 10$	$7 - 15$	$0.1 - 0.7$

2.2 Thermal Properties of Star-forming Clouds

Molecular clouds are cold (Cernicharo, 1991), see Section 2.1. The kinetic temperature inferred from molecular line ratios is typically about 10 K for dark, quiescent clouds and dense cores in GMCs that are shielded from UV radiation by high column densities of dust. Nevertheless it can reach $50 - 100$ K in regions heated from the inside by UV radiation from high-mass stars. For example, the temperature of gas and dust behind the Trapezium cluster in Orion is about 50 K. The thermal structure of the gas is related to its density distribution and its chemical composition, so it is remarkable that over a wide range of gas densities and metallicities the equilibrium temperature remains almost constant in a small range around $T \approx 10$ K (Goldsmith & Langer, 1978; Goldsmith, 2001). These observational results influence the theoretical treatment of the thermal properties of star-forming gas.

Early studies of the balance between heating and cooling processes in collapsing clouds predicted temperatures of the order of 10 K to 20 K, tending to be lower at higher densities (e.g. Hayashi & Nakano, 1965; Hayashi, 1966; Larson, 1969, 1973b). In their dynamical collapse calculations, these and other authors approximated this varying temperature by a simple constant value, usually taken to be 10 K. Nearly all subsequent studies of cloud collapse and fragmentation have used a similar isothermal approximation. However, this

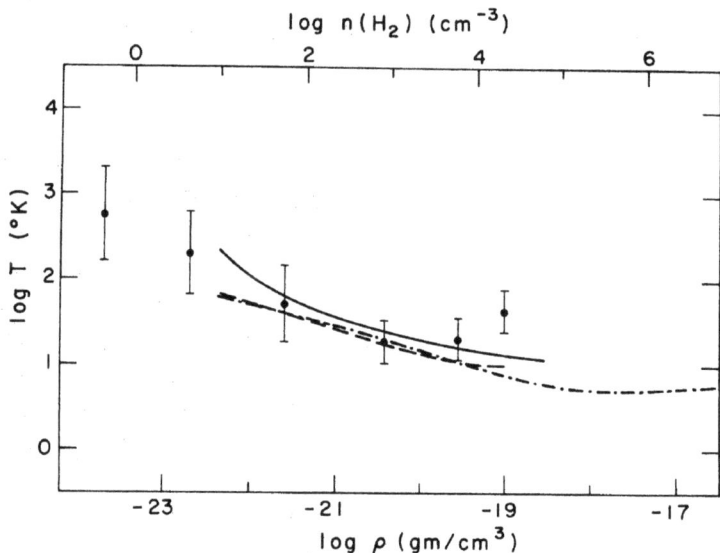

Figure 2.4: Theoretical and observed temperatures of interstellar clouds as a function of density. Dots with error bars are the logarithmic averages and standard deviations of the measured cloud temperatures compiled by Myers (1978); the solid curve is the semi-empirical temperature-density relation of Tarafdar et al. (1985); the dashed curve is the theoretical relation of Boland & de Jong (1984); and the dot-dash curve is the earlier theoretical relation of Larson (1973b). From Larson (1985).

approximation is actually only a somewhat crude one, valid only to a factor of 2, since the temperature is predicted to vary by this much above and below the usually assumed constant value of 10 K. Given the strong sensitivity of the results of fragmentation simulations (see, e.g. Li et al., 2003) to the assumed equation of state of the gas, temperature variations of this magnitude may be important for quantitative predictions of stellar masses and the initial stellar mass function (see Section 2.3).

As can be seen in Figure 2.4, both, observational and theoretical studies of the thermal properties of collapsing clouds indicate that at gas mass densities below about $10^{-18}\,\mathrm{g\,cm^{-3}}$, roughly corresponding to a gas number density of $n = 2.5 \times 10^5\,\mathrm{cm^{-3}}$, the temperature generally decreases with increasing density. In this relatively low-density regime, clouds are externally heated by cosmic rays or photoelectric heating, and they are cooled mainly by the collisional excitation of low-lying levels of C^+ ions and O atoms; the strong dependence of the cooling rate on density then yields an equilibrium temperature that decreases with increasing density. The work of Koyama & Inutsuka (2000), which assumes that photoelectric heating dominates, rather than cosmic ray heating as had been assumed in earlier work, predicts a very similar trend of decreasing temperature with increasing density at low densities. The resulting gas temperature-density relation can be approximated by a power law with an exponent of about -0.275, which corresponds to a polytropic equation of state with $\gamma = 0.725$. The observational results of Myers (1978), shown in Fig. 2 of Larson (1985), suggest temperatures rising again toward the high end of this low-density regime, but those measurements refer mainly to relatively massive and warm cloud cores and not to the small, dense, cold cores in which low-mass stars form. As reviewed by Evans (1999), the temperatures of these cores are typically only about 8.5 K at a density of $10^{-18}\,\mathrm{g\,cm^{-3}}$, consistent with a continuation of the decreasing trend noted above and with the continuing validity of a polytropic approximation with $\gamma \approx 0.725$ as low a density as at least $10^{-18}\,\mathrm{g\,cm^{-3}}$.

At densities higher than this, star-forming cloud cores become opaque to the heating and cooling radiation that determines their temperatures at lower densities, and at densities above $10^{-18}\,\mathrm{g\,cm^{-3}}$ the gas becomes thermally coupled to the dust grains, which then control the temperature by their far-infrared thermal emission. In this high-density regime, dominated thermally by the dust, there are few direct temperature measurements because the molecules normally observed freeze out onto the dust grains, but most of the available theoretical predictions are in good agreement concerning the expected thermal behavior of the gas (Larson, 1973b; Low & Lynden-Bell, 1976; Masunaga & Inutsuka, 2000). The balance between compressional heating and thermal cooling by dust results in a temperature that increases slowly with increasing density, and the resulting temperature-density relation can be approximated by a power law with an exponent of about 0.075, which corresponds to $\gamma = 1.075$. Taking these values, the temperature is predicted to reach a minimum of 5 K at the transition between the low-density and the high-density regime at about $2\times 10^{-18}\,\mathrm{g\,cm^{-3}}$, at which point the Jeans mass is about $0.3\,M_\odot$ (see also, Larson, 2005). The actual minimum temperature reached is somewhat uncertain because observations have not yet confirmed the predicted very low values, but such cold gas would be very difficult to observe; various efforts to model the observations have suggested central temperatures between 6 K and 10 K for the densest observed prestellar cores, whose peak densities may approach $10^{-17}\,\mathrm{g\,cm^{-3}}$ (e.g. Zucconi et al., 2001; Evans et al., 2001; Tafalla et al., 2004). A power-law approximation to the equation of state with $\gamma \approx 1.075$ is expected to remain valid up to a density of about $10^{-13}\,\mathrm{g\,cm^{-3}}$, above which increasing opacity to the thermal emission from dust causes the temperature to begin rising much more rapidly, resulting in an "opacity limit" on fragmentation that is somewhat below $0.01\,M_\odot$ (Low & Lynden-Bell, 1976; Masunaga & Inutsuka, 2000).

2.3 The IMF

One of the fundamental unsolved problems in astronomy is the origin of the stellar mass spectrum, the so-called initial mass function (IMF). The IMF gives the relative number of stars formed per unit mass interval, i.e. the distribution of stellar masses at birth. The Mass Function (MF) was originally defined by Salpeter (1955) as the number of stars N in a volume of space V observed at a time t per logarithmic mass interval $d\log m$:

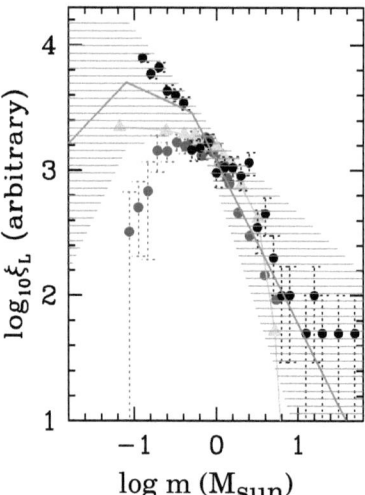

Figure 2.5: The measured stellar mass function ξ_L as a function of logarithmic mass $\log m$ in the Orion nebular cluster (upper circles), the Pleiades (triangles connected by line), and the cluster M35 (lower circles). None of the mass functions is corrected for unresolved multiple stellar systems. The average initial stellar mass function derived from Galactic field stars in the solar neighborhood is shown as a line with the associated uncertainty range indicated by the hatched area. From Kroupa (2002).

$$\xi_L(\log m) = \frac{d(N/V)}{d \log m} = \frac{dn}{d \log m}, \qquad (2.1)$$

where $n = N/V$ is the stellar number density, which is in pc^{-3} in the following. Conversely Scalo (1986) defines the mass spectrum as the number density distribution per mass interval dn/dm with the relation

$$\xi(m) = \frac{dn}{dm} = \frac{1}{m}\frac{dn}{d \log m} = \frac{1}{m}\xi_L(\log m). \qquad (2.2)$$

With these definitions, if the MF is approximated as a power law, the exponents are usually denoted, respectively, x and α, with $\xi_L(\log m) \propto m^{-x}$ and $\xi(m) \propto m^{-\alpha}$, $x = \alpha - 1$.

Stars evolve off the main sequence (MS) after a certain age, so the present-day mass function (PDMF) of MS stars, which can be determined from the observed present-day luminosity function, differs from the IMF as defined above. As noted by Miller & Scalo (1979), all stars with MS lifetimes greater than the age of the Galaxy are still on the MS.

In that case the PDMF and the IMF are equivalent. For stars with MS lifetimes less than the age of the Galaxy, i.e. massive stars, only those formed within the last MS lifetime are observed today as MS stars. In that case, the PDMF and the IMF are different (Chabrier, 2003).

It is complicated and laborious to estimate the IMF in our Galaxy empirically. The first such determination from the solar neighborhood (Salpeter, 1955) showed that IMF can be approximated by a power law with index $x \approx 1.35$ or $\alpha \approx 2.35$ for stars in the mass range $0.4 \leq m \leq 10$. However, approximation of the IMF with a single power law is too simple. Miller & Scalo (1979) introduced a log-normal functional form, again to describe the IMF for the Galactic field stars in the vicinity of the Sun,

$$\log \xi_L(\log m) = A - \frac{1}{2(\log \sigma)^2}(\log(m/m_0))^2. \qquad (2.3)$$

This analysis has been repeated and improved upon by Kroupa et al. (1990), who derive values: $m_0 = 0.23\,M_\odot$, $\sigma = 0.42$, $A = 0.1$. However, both studies did not take into account the contamination of star counts by binary and multiple systems.

The IMF can also be estimated, probably more directly, by studying individual young star clusters. Typical examples are given in Figure 2.5 (taken from Kroupa 2002), which plots the mass function derived from star counts in the Trapezium Cluster in Orion (Hillenbrand & Carpenter, 2000), in the Pleiades (Hambly et al., 1999), and in the cluster M35 (Barrado y Navascués et al., 2001).

The most popular approach to approximating the IMF empirically based on Galactic field stars is to use a multiple-component power law with the following parameters (Scalo, 1998; Kroupa, 2002):

$$\xi(m) = \begin{cases} 0.26\,m^{-0.3} & \text{for} \quad 0.01 \leq m < 0.08 \\ 0.035\,m^{-1.3} & \text{for} \quad 0.08 \leq m < 0.5 \\ 0.019\,m^{-2.3} & \text{for} \quad 0.5 \leq m < \infty \end{cases}. \qquad (2.4)$$

The IMF may steepen further towards high stellar masses and a fourth component could be defined with $\xi(m) = 0.019\,m^{-2.7}$ for $m > 1.0$ (Kroupa et al., 1993; Chabrier, 2003). In Equation 2.4, the exponents for masses $m < 0.5\,M_\odot$ are very uncertain due to the difficulty of detecting and determining the masses of very young low-mass stars. The exponent for $0.08 \leq m < 0.5$ could vary between -0.7 and -1.8, and the value in the substellar regime is even less certain. The above representation of the IMF which is defined for single stars is statistically corrected for binary and multiple stellar systems too close to be resolved, but too far apart to be detected spectroscopically. Neglecting these systems overestimates the number of high-mass stars, while underestimating the number of low-mass stars.

Despite the differences between the IMF obtained from field stars and the IMF of young star clusters, all IMF determinations share the same basic features, and it appears reasonable to say that the basic shape of the IMF is a universal property common to all star-forming regions in the present-day Galaxy, perhaps with some intrinsic scatter (Kroupa, 2001). However, there may still be some dependency on the metallicity of the star-forming gas. In addition, the stellar mass function in the Arches cluster (Stolte et al., 2005) near the center of our Galaxy is found to have a deficit of stars below 7 $M_\odot$. This gives observational support to the notion that the IMF shows qualitatively different behavior in regions with special conditions, such as the Galactic center or starburst environments (Klessen et al., 2007)

The IMF has a nearly power-law form for larger masses and declines rapidly towards smaller masses (Scalo, 1998; Kroupa, 2002; Chabrier, 2003). The IMF thus has a peak at a characteristic mass of typically a few tenths of a solar mass in star-forming regions in the solar vicinity. In contrast, the initial conditions in star forming regions can vary considerably. If the IMF depends on the initial conditions, there would thus be no reason for it to be universal. Therefore a derivation of the characteristic stellar mass that is based on fundamental atomic and molecular physics would be more consistent. In this work we make a first step towards describing the thermal properties of the star-forming gas more precisely (see Section 3.4.2).

There have been analytical models (e.g. Jeans, 1902; Larson, 1969; Penston, 1969a; Low & Lynden-Bell, 1976; Shu, 1977; Whitworth & Summers, 1985) and numerical investigations of the effects of various physical processes on collapse and fragmentation. These processes include, for example, magnetic fields (Basu & Mouschovias, 1995; Tomisaka, 1996; Galli et al., 2001), feedback from the stars themselves (Silk, 1995; Nakano et al., 1995; Adams & Fatuzzo, 1996) and competitive coagulation or accretion (Silk & Takahashi, 1979; Lejeune & Bastien, 1986; Price & Podsiadlowski, 1995; Murray & Lin, 1996; Bonnell et al., 2001a,b; Durisen et al., 2001). In another group of models, initial and environmental conditions, like the structural properties of molecular clouds, determine the IMF (Elmegreen & Mathieu, 1983; Elmegreen, 1997a,b, 1999, 2000c,a, 2002). Larson (1973a) and Zinnecker (1984, 1990) argued in a more statistical approach that the central-limit theorem naturally leads to a log-normal stellar mass spectrum at the low-mass end. Moreover, there are models that connect turbulent motions in molecular clouds to the IMF (e.g. Larson, 1981; Fleck, 1982; Padoan, 1995; Padoan et al., 1997; Klessen et al., 1998, 2000; Klessen, 2001; Padoan & Nordlund, 2002; Ballesteros-Paredes et al., 2006). Recently, Elmegreen & Scalo (2006) point out the importance of the star formation history for the shape of the IMF. Knowledge of the distribution of stellar masses at birth, described by the IMF, is necessary to understand many astrophysical phenomena, but no analytic derivation of the observed IMF has yet explained all observations consistently. In fact, it appears likely that a fully deterministic theory for the IMF does not exist. Rather, any viable theory must take into account the probabilistic nature of the turbulent process of star formation, which is inevitably highly stochastic and indeterminate.

Current attempts to understand the nature of the IMF, as those mentioned above, are generally based on models that do not treat the thermal physics in detail. Typically, they use a simple isothermal equation of state (see Section 3.4.1). In this thesis we make an attempt to diverge from this method by modeling the thermal properties of molecular clouds with a piecewise polytropic equation of state as explained in detail in Section 3.4.2. In the long run, one would like to take the chemical, cooling and heating as well as radiative processes into account. Our next step therefore is to include time-dependent chemistry into our dynamical model and we focus on a special case: the collapse of initially ionized gas towards the formation of the second generation of stars in the universe. In protogalactic halos the gas has only low metal content and the cooling processes are mainly dominated by H_2. In

Section 4.5 we describe in detail how cooling and the chemistry in this environment can be modeled and what information on the thermal behavior of the gas we can infer.

2.4 The Angular Momentum Problem

Gravitational collapse in the context of star formation always involves solving the angular momentum problem. This results from the discrepancy between the specific angular momentum observed in low-density gas on large scales and the amount of rotation or orbital angular momentum present after collapse (Spitzer, 1968; Bodenheimer, 1995). The source of angular momentum on large scales lies in the rotation of the galactic disk and, closely related to that, on intermediate to small scales it results from the high degree of vorticity inextricably inherent in turbulent flows. The typical specific angular momentum j of molecular cloud material, e.g. on scales of about 1 pc, is $j \approx 10^{23}\,\mathrm{cm^2\,s^{-1}}$, while on scales of cloud cores, say below 0.1 pc, it is of order $10^{21}\,\mathrm{cm^2\,s^{-1}}$. A 1 $M_\odot$ binary star with an orbital period of 3 days has $j \approx 10^{19}\,\mathrm{cm^2\,s^{-1}}$, while the spin of a typical T Tauri star is a few $\times 10^{17}\,\mathrm{cm^2\,s^{-1}}$. Our own Sun has only a specific angular momentum of $j \approx 10^{15}\,\mathrm{cm^2\,s^{-1}}$. This means that during the process of star formation most of the initial angular momentum is removed from the collapsing object.

The presence of magnetic fields provides, in principle, a viable mechanism for locally reducing the angular momentum through magnetic braking. This was treated approximately by Ebert, von Hörner & Temesváry (1960), and later calculated accurately by Mouschovias & Paleologo (1979, 1980). The criterion for effective braking is essentially that helical Alfvén waves excited by the rotating cloud have to couple to the ambient medium over a volume that contains roughly the same mass as the cloud itself. For the strong magnetic fields required by the standard theory of star formation, the deceleration time can be less than the free-fall time, leading to efficient transfer of angular momentum away from collapsing cores, and thus, to the formation of single stars.

Table 2.2: Characteristic values of specific angular momentum

Cloud Type		Specific Angular Momentum [cm^{2}s^{-1}]
Molecular Cloud	(scale: 1 pc)	10^{23}
Molecular Cloud Core	(scale: 0.1 pc)	10^{21}
Binary	(period: 10^4 yr)	$4 \times 10^{20} - 10^{21}$
Binary	(period: 10 yr)	$4 \times 10^{19} - 10^{20}$
Binary	(period: 10 yr)	$4 \times 10^{18} - 10^{19}$
Disk around 1 $M_\odot$ central star	(radius: 100 AU)	4.5×10^{20}
T Tauri star	(spin)	5×10^{17}
Jupiter	(orbit)	10^{20}
Sun	(present spin)	10^{15}

Table is adapted from Bodenheimer (1995).

It is therefore a crucial test for any theory of star formation whether it can produce the required angular momentum loss during collapse while at the same time explain the high numbers of binaries and multiple stellar systems observed (e.g. Duquennoy & Mayor, 1991; Halbwachs et al., 2003). In a semi-empirical analysis of isolated binary star formation Fisher (2004) presented the effects of non-magnetical turbulence in the initial state of the gas on binary orbital parameters. These properties were in agreement with observations if a significant loss of angular momentum was assumed. In the current investigation we focus on numerical simulations of non-magnetic, supersonically turbulent, self-gravitating clouds and analyze the time evolution of angular momentum during formation and subsequent collapse of protostellar cores. Our main question is whether gravoturbulent fragmentation can solve or at least ease the so called "angular momentum" problem without invoking the presence of magnetic fields. We present the simulations that address this issue and our results in Chapter 5.

Chapter 3

Modeling Star Formation in Different Epochs

3.1 Cosmological Preliminaries

3.1.1 The Standard Model

Besides present-day star formation, we also address the question of early and primordial star formation in the thesis (see Chapter 7). In the latter case, we need to place the discussion in its proper cosmological context. In this section we will give a short introduction into the most important aspects of the early cosmic evolution.

Following Peebles (1993), we can identify a 'standard model' of physical cosmology – a world picture which, although incomplete, has passed many observational tests and is consistent with the available evidence. This cosmological principle asserts that the distribution of matter in the universe is homogeneous and isotropic if smoothed on a sufficiently large scale, that it is expanding and that it was substantially hotter and denser in the past. It is based upon two assumptions and a number of important observations.

The first assumption is simply that the laws of physics on cosmological scales are the same as those on smaller scales; in particular, that general relativity is the correct description of gravity on large scales. This assumption is certainly consistent with the evidence, but it is wise to remember that it involves an extrapolation from the terrestrial scale on which general relativity is well tested to a scale that is many orders of magnitude larger. The other assumption that we make is often termed the Copernican Principle: the assumption that there is nothing particularly special about the position we occupy in the universe. Moreover, there appears to be nothing particularly special about the Milky Way compared with, say, M31 or M101, and there seems to be no good reason to suppose that our presence in this particular galaxy is due to anything other than chance.

Turning to the observations, we find a mixture of direct and indirect evidence for the model. The cosmic microwave background (CMB) and the hard X-ray background both provide strong evidence of isotropy (Wu et al., 1999). Moreover, this is also strong evidence for homogeneity; the observed isotropy is a natural consequence of a homogeneous universe, but it would require us to be in a particularly special location in an inhomogeneous universe. We can also test homogeneity directly, by means of galaxy counts.

On small scales, galaxies are known to cluster inhomogeneously, but if we smooth the number counts on larger and larger scales, we expect to see a transition to homogeneity. Various large redshift surveys that are currently underway, such as the 2degreeField (2dF, Maddox, 2000) or the Sloan Digital Sky Survey (York et al., 2000), have shown that voids in the distribution of galaxies are not larger than $\sim 100\,\mathrm{Mpc}$; on larger scales there seems to be evidence for homogeneity. Evidence for expansion comes from the redshift-distance relation: the fact that the spectra of extragalactic sources are shifted to the red by an amount that is directly proportional to distance d. This was first noted by Wirtz (1924), but was first quantified by Hubble (1929), and has since become known as Hubble's law:

$$v_\mathrm{r} = \frac{\lambda_\mathrm{observed} - \lambda_\mathrm{emitted}}{\lambda_\mathrm{emitted}} c = z \cdot c = H_0 \cdot d, \tag{3.1}$$

where v_r is the radial velocity of the source, $\lambda_\mathrm{observed}$ the observed wavelength, λ_emitted the emitted wavelength and c the speed of light. The proportionality constant H_0 is known as the Hubble parameter and z is known as the redshift of the object. Hubble's law is a natural consequence of an expanding universe, as long as the expansion is homogeneous. Another natural consequence of an expanding universe, as long as the energy is conserved, is the assertion that it has been hotter and denser in the past. The evidence that this was indeed the case comes from two different sources: the observed elemental abundances and the microwave background.

By observing elemental abundances in regions that have undergone little or no star formation (Adams, 1976), we can attempt to infer the primordial abundances of the various elements. We find that the inferred abundances are consistent with the predictions of the expanding universe model, to within the observational uncertainties (Burles et al., 2001). Moreover, the theoretical model of primordial nucleosynthesis has only one free parameter – the baryon-photon ratio.

The microwave background is also strong evidence for a hot, dense phase of the early universe. In particular, measurements by the FIRAS experiment on the COBE satellite have shown that the CMB has a spectrum which corresponds to an almost perfect black-body; any deviation is at the level of one part in ten thousand or less (Fixsen et al., 1996). This observation is simple to explain in the standard model. At an early epoch, the temperature is high enough that all of the gas is ionized and consequently radiation and matter are closely coupled by Thomson scattering and in thermal equilibrium. At this time, the radiation spectrum is necessarily that of a black-body. As the universe expands, however, the temperature drops, and the gas eventually begins to recombine. At this point, the Thomson scattering optical depth drops sharply. The bulk of the radiation subsequently never interacts with matter, and thus, aside from the effects of redshift[1], the spectrum remains unchanged to the

---

[1]Redshifting a black-body spectrum does not alter its shape, but merely lowers its characteristic temper-

present day. To produce a black-body CMB from other sources of radiation would be far harder. It would require the action of some mechanism capable of thermalizing the spectrum, but at the same time having no discernable effects on the spectra of extragalactic sources visible at the present day. Increasingly precise measurements of the cosmic microwave background (CMB), as exemplified by the recent results of wmap (Bennett et al., 2003), have helped us confirm that we live in a flat universe, with approximately 5% of the closure density provided by baryons, 25% by cold dark matter (CDM), and the remaining 70% by some form of 'dark energy' or cosmological constant. Models of such a universe – generally known as ΛCDM models – have been heavily studied for a number of years and many of their features are well understood. For instance, the evolution of the small inhomogeneities in the early universe that give rise to the observed temperature anisotropies in the CMB can be followed in great detail (Seljak & Zaldarriaga, 1996), and the resulting predictions have been strongly confirmed by the wmap results.

The evolution of the dark matter component of the universe subsequent to the epoch of last scattering at $z \simeq 1100$ has also been studied intensively, using a wide range of techniques (see for example Seljak, 2000; Benson et al., 2001; Cooray & Sheth, 2002). The general agreement between the results of these studies and an increasing number of observational tests (e.g. Gray et al. 2002) has lent further support to this overall picture, although some puzzles remain (Moore et al., 1999; Navarro & Steinmetz, 2000).

Individually, none of these pieces of evidence are entirely persuasive – if we tried hard enough we could usually construct some model to explain them. Taken together, however, the fact that the same simple model explains all of the otherwise unrelated observational evidence argues strongly for its basic correctness. Thus, it provides us with a basic framework within which our other theories must fit.

In our simulations we adopt the cosmological parameters taken from the wmap concordance model (Spergel et al., 2003). Specifically: matter density $\Omega_m = 0.29$, cosmological constant (energy density of the vacuum) $\Omega_\Lambda = 0.71$, baryon density $\Omega_b = 0.047$, Hubble constant $h = 0.72$, amplitude of galaxy fluctuations $\sigma_8 = 0.9$ and spectral index $n_s = 0.99$.

3.1.2 Hierarchical Structure Formation

The standard model is a good description of the universe on scales on which we can regard it as effectively homogeneous. However, when we look at the universe on smaller scales, we find that it is largely inhomogeneous. Galaxies have mean densities that are orders of magnitude larger than the cosmological background density, while, on larger scales, groups, clusters and superclusters also demonstrate the existence of departures from homogeneity. Clearly, an obvious question to ask is how this wealth of structure arises in an initially homogeneous universe.

We know from observations that the cosmic background radiation shows small variations in intensity between two different directions. If we interpret the universe as a black-body and take the differences in intensities to represent temperature fluctuations, then the maximum

---

ature.

difference observed is $\Delta T \sim 10^{-5}$ K. As it might be expected, the study of how these small perturbations develop into the structure that we see around us today forms a major branch of cosmology.

The origin of the seed fluctuations will not be discussed here, we confine ourselves to a discussion on the growth of these fluctuations into the clusters of galaxies that we see today. A seed fluctuation is a region of the universe which had a density greater than the average density of the universe. This overdensity is given by

$$\Delta = \frac{\rho - \bar{\rho}}{\bar{\rho}}$$

where ρ is the density of the region and $\bar{\rho}$ is the average density of the universe. Two competing effects act on the seed fluctuations: self-gravity, causing them to grow in mass and density, and the expansion of the universe, dispersing the fluctuations and decreasing their density. Because the overall density of the universe is decreasing, the fluctuations will grow in overdensity even if their actual density is decreasing until collapse takes over. This will happen when $\Delta \sim 1$, i.e. when the density of the fluctuation is roughly twice the average density. Note that the growth of baryonic fluctuations can only occur after the decoupling of matter and radiation, before which the scattering of photons off ionized gas particles tends to smooth out any matter density fluctuations.

Historically, there are two competing theories for how structures continued to evolve after collapse. Either the first structures to collapse were very large, on the scale of clusters of galaxies, which then fragmented and formed galaxies, or structures first appeared on the scale of individual galaxies which progressed to form groups and clusters of galaxies through hierarchical merging (Peebles, 1993). In CDM models, gravitationally bound objects form in the latter way. Structure formation of the dark matter proceeds in a hierarchical, or 'bottom-up' fashion, with the smallest, least massive objects forming first, and larger objects forming later through a mixture of mergers and accretion. The most compelling evidence for this mechanism of structure formation are observations of galaxies existing at very high redshifts (e.g. Becker et al., 2001). Although there are very few clusters known at redshifts greater than 1, there is evidence that the most massive clusters have formed at about $z = 1$ (e.g. Henry et al., 1992). The subsequent formation of larger objects occurs rapidly, and at most redshifts a large number of gravitationally bound objects (frequently referred to as 'dark matter halos') exists, with a wide range of masses. Considerable effort has been devoted to determining the mass function of dark matter halos as a function of redshift. The most widely used expression for the mass function is the one originally suggested by Press & Schechter (1974):

$$n(M, z)\,\mathrm{d}M = \sqrt{\frac{2}{\pi}} \frac{\rho_{\mathrm{dm}}}{M} \frac{\mathrm{d}\nu}{\mathrm{d}M} \exp\left(-\frac{\nu^2}{2}\right) \mathrm{d}M. \tag{3.2}$$

Here $n(M, z)\,\mathrm{d}M$ is the comoving number density of halos at redshift z with dark matter halo masses in the interval $(M, M + \mathrm{d}M)$, ρ_{dm} is the cosmological background dark matter density, and $\nu \equiv \delta_{\mathrm{c}}/[D(z)\sigma(M)]$, where δ_{c} is a critical overdensity (generally taken to be 1.69), $D(z)$ is the linear growth factor (Peebles, 1980; Carroll et al., 1992) and $\sigma(M)$ is the root-mean-square (rms) fluctuation in the cosmological dark matter density field smoothed on a mass scale M. A comprehensive discussion of the derivation of this equation is given in Bond et al. (1991).

In CDM models, $\sigma(M)$ decreases monotonically with increasing mass, and so the most massive objects will also be the rarest. The transition to exponential behavior occurs for $\nu \sim 1$, or $\sigma(M) \sim \delta_c/D(z)$, and so this transition occurs at a progressively smaller mass as we move to higher redshifts.

Given a mass function of this type, is there any way to specify when the first halo of a given mass forms? Strictly speaking, the answer is no; the probability of finding a halo of any finite mass is never zero. In practice, however, we are more interested in determining when this probability grows to some interesting size, or when the number density of halos exceeds some specified threshold. This is most commonly calculated by specifying a value of ν which is of interest; for instance, reference is often made to 3σ halos, which are simply halos for which $\nu = 3$ and which therefore have a dark matter mass M satisfying:

$$\sigma(M) = \frac{1}{3}\frac{\delta_c}{D(z)}. \quad (3.3)$$

Such halos are moderately rare objects, representing no more than a few thousandths of the total cosmic mass (Mo & White, 2002), but are sufficiently common, so that one would expect to find many of them within a single Hubble volume. They are often taken to be representative of the earliest objects to form, although this choice is somewhat arbitrary.

Unfortunately, while the Press-Schechter approach allows us to determine when the first dark matter halos of a given mass form, it does not, by itself, tell us when the first protogalaxies form, as it contains no information about the behavior of the baryonic component of the universe. Unlike the dark matter, the baryons do not initially form structures on very small scales, since pressure forces act to suppress the growth of small-scale perturbations (Jeans, 1902; Bonnor, 1957). Using linear perturbation theory Peebles (1980) shows that in a purely baryonic universe, the growth of perturbations is completely suppressed on scales smaller than

$$\lambda_J \leq \frac{\pi^{1/2} c_s}{\sqrt{G \rho_b}}, \quad (3.4)$$

where ρ_b is the cosmological baryon density. This critical wavelength is commonly known as the Jeans length (see Section 3.3.2). The associated mass scale, known as the Jeans mass is defined in Equation 3.15 (where ρ_0 should be replaced by ρ_b). The value of the Jeans mass depends on the baryon density, which is a simple function of redshift, and on the temperature of the intergalactic medium (through the dependence of λ_J on c_s). The latter is simple to calculate at epochs prior to the onset of widespread star formation and is well approximated by Galli & Palla (1998)

$$T = 410 \left(\frac{1+z}{150}\right)^2 \text{ K} \quad (3.5)$$

for redshifts $z < 150$. The corresponding Jeans mass at these redshifts is given by

$$M_J = \frac{4.9 \times 10^4}{(\Omega_b h^2)^{1/2}} \left(\frac{1+z}{150}\right)^{3/2} M_\odot. \quad (3.6)$$

To generalize this to the case of a universe containing both baryons and cold dark matter, one can replace the baryon density in the above equations with the total density $\rho_m = \rho_b + \rho_{dm}$, which would give us

$$M_J = \frac{4.9 \times 10^4}{(\Omega_m h^2)^{1/2}} \left(\frac{1+z}{150}\right)^{3/2} M_\odot \qquad (3.7)$$

for $z < 150$; or in other words, a Jeans mass that is a factor $(\Omega_b/\Omega_m)^{1/2}$ smaller. Using Equation 3.7, Glover (2005) estimates that protogalaxies will develop within 3σ halos once the mass of the dark matter in the halo exceeds M_J which occurs at $z \sim 30$.

3.2 Star Formation Theory - Two Short Histories

3.2.1 Early Star Formation

In the preceding section we discussed the hierarchical structure formation within the framework of a cold dark matter (CDM) cosmology. In this section we show how this influences the formation of the first and second generation of stars. We also give a short overview of the relevant theoretical work.

At high redshifts, CDM-dominated low-mass halos with virial temperatures less than $\approx 10^4$ K are abundant. Primordial gas in these halos cools by molecular hydrogen transitions, because H_2 is the only coolant present in significant quantities that remains effective at temperatures below 10^4 K (Saslaw & Zipoy, 1967; Peebles & Dicke, 1968; Matsuda et al., 1969). Tegmark et al. (1997) developed analytic methods to model early baryonic collapse via H_2 cooling. Numerical studies of the formation of primordial gas clouds and the first stars indicate that this process likely began as early as $z \sim 30$ (Abel et al., 2002; Bromm et al., 2002). Yoshida et al. (2003) further utilized simulations to develop a semi-analytic model based on the Tegmark et al. (1997) method and included the effects of dynamical heating caused by the thermalization of kinetic energy of infall into a deepening potential. Both approaches suggest that only gas in halos more massive than some critical mass M_{crit} will cool efficiently (reviewed recently by Bromm & Larson, 2004; Ciardi & Ferrara, 2005; Glover, 2005).

Population III stars were the first potential sources of UV photons that contribute to the reionization process, and produced most of the metals required for the formation of population II stars. An important question is whether later generations of stars can efficiently form in the relatively high temperatures and ionization fractions of the relic H ii regions left by the first stars. One analytical study (Oh & Haiman, 2003) found that the first stars injected sufficient energy into the early intergalactic material (IGM), by means of photoheating and supernova explosions, to prevent further local star formation in their vicinity. The Lyman-Werner UV background is also thought to have contributed negative feedback by photodissociating primordial H_2 and quenching the molecular hydrogen cooling processes, thereby delaying cooling and collapse of the primordial gas (Haiman et al., 2000; Machacek et al., 2001). Metals produced by the first stars were also be injected into the gas in protogalactic halos that have not collapsed yet. The question that then arises is how this metallicity affected the ability of the gas to cool and collapse. Bromm et al. (2001) argued that there exists a critical metallicity $Z_{\text{crit}} = 5 \times 10^{-4}$ below which the gas fails to undergo continued collapse and fragmentation. Nevertheless in their simulations they did not take H_2 cooling into account although this is still present if self-shielding is effective.

3.2.2 Current Star Formation

In the universe stars form from gravitational contraction of gas and dust in molecular clouds. This fact is at the center of all attempts to combine the observational results with a theory of star formation. In particular the pioneering work by Jeans (1902) concerning the importance of gravitational instability for stellar birth (see Section 3.3.2) triggered numerous attempts to derive solutions to the collapse problem, both analytically and numerically (e.g. Bonnor, 1956; Ebert, 1957; Larson, 1969; Penston, 1969b, for some early studies). The classical dynamical theory focuses on the interplay between self-gravity and pressure gradients. Turbulence is taken into account only on microscopic scales significantly smaller than the collapse scales. In this microturbulent regime (Chandrasekhar, 1951b,a), random gas motions yield an isotropic pressure that can be absorbed into the equation of motion as an effective sound speed (see Section 3.3.3).

From observations it became clear that substantial magnetic fields thread the interstellar medium (Chandrasekhar & Fermi, 1953a). This raised the possibility that the solution to the angular momentum problem might be found in the action of magnetic fields. Magnetic braking provides a mechanism that results in the loss of angular momentum during collapse (e.g. Basu & Mouschovias, 1994). Moreover, there exists a magnetic flux problem as stars are left with little magnetic flux compared to the star-forming clouds. This problem was addressed in the model of low-mass star formation of Shu et al. (1987). It involves the collapse of an isothermal sphere producing a single protostar. The isothermal sphere loses its magnetic support via ambipolar diffusion, where neutrals slowly decouple from the ions and the magnetic field to produce a $\rho \propto r^{-2}$ density profile. Collapse of a sphere with such a profile gives a constant accretion rate. However, Whitworth et al. (1996) have argued that such a density profile is unlikely to arise in nature. In addition, Class 0 protostars are observed to be undergoing collapse with a less centrally condensed profile (Ward-Thompson et al., 1994). More accurate determinations of magnetic field strengths in molecular clouds showed that cloud cores are not necessarily supported by the magnetic field (Bourke et al., 2001). Recently the concept of gravoturbulent fragmentation, which we will introduce in the next section, has reconciled the theoretical results with the observational evidence.

3.3 Gravoturbulent Fragmentation

In the center of our model of star formation stands the notion of gravoturbulent fragmentation (e.g. Mac Low & Klessen, 2004). It is based on the interplay between turbulent motions in gas clouds and the self-gravity of the gas cloud. As cited in Chapter 2, observations are consistent with the idea that star-forming clouds are supported by supersonic turbulence (e.g. molecular emission lines show signs of supersonic motions). Supersonic turbulence produces strong density fluctuations in the interstellar gas, sweeping up gas from large regions into dense sheets and filaments, and does so even in the presence of magnetic fields (e.g. Vázquez-Semadeni et al., 2000; Heitsch et al., 2001). Supersonic turbulence decays quickly (Gammie & Ostriker, 1996; Stone et al., 1998; Mac Low et al., 1998), but so long as it is maintained by input of energy from some driver, it can support large-scale regions against gravitational collapse (e.g. Mac Low, 1999; Klessen et al., 2000). Such support comes at a cost, however. The same turbulent flows that support a region globally produce local density enhancements with gravity taking over in the densest and most massive parts. Once gas clumps become gravitationally unstable, collapse sets in. In the following sections we discuss how the gas and the above mentioned processes that lead to star formation can be described theoretically. We will also introduce the concepts necessary to the theory of gravoturbulence. For further details see Mac Low & Klessen (2004).

3.3.1 Self-Gravitating Hydrodynamics

The gas in the interstellar medium is highly compressible and it is subject to its own self-gravity. A simple description may be obtained by modeling the star-forming molecular gas as an ideal, inviscid, self-gravitating, non-magnetic gas. The state of the gas is determined by four parameters, namely the velocity $\vec{v}$, the pressure P, the specific internal energy u, and the density ρ, which are functions of position $\vec{r}$ and time t. The gas is governed by the following equations

$$\frac{d\rho}{dt} = \frac{\partial \rho}{\partial t} + \vec{v} \cdot \vec{\nabla} \rho = -\rho \vec{\nabla} \cdot \vec{v} \tag{3.8}$$

$$\frac{d\vec{v}}{dt} = \frac{\partial \vec{v}}{\partial t} + (\vec{v} \cdot \vec{\nabla})\vec{v} = -\frac{\vec{\nabla} P}{\rho} - \vec{\nabla}\Phi \tag{3.9}$$

$$\frac{du}{dt} = \frac{\partial u}{\partial t} + \vec{v} \cdot \vec{\nabla} u = -\frac{P}{\rho}\vec{\nabla} \cdot \vec{v} - \frac{\Lambda(u,\rho)}{\rho} \tag{3.10}$$

$$\Delta \Phi = 4\pi G \rho \tag{3.11}$$

$$P = \mathcal{R}\rho T, \tag{3.12}$$

where Φ denotes the gravitational potential.

The continuity Equation 3.8 expresses the conservation of mass. Nevertheless it implies more than simple mass conservation, for it states that changes in the local matter content due to fluid flow occur in a continuous fashion. Mass loss from any volume element will occur by matter flowing in a well-defined manner across the surface of the volume.

Equation 3.9 which is also called Euler's equation reflects the conservation of momentum, where $-\vec{\nabla}\Phi$ represents the acceleration induced by the gravitational potential Φ which is obtained from the density distribution via Poisson's Equation 3.11. We can also add a viscous acceleration $(\vec{\nabla} \cdot \overleftrightarrow{\pi})/\rho$, where $\overleftrightarrow{\pi}$ denotes the viscous stress tensor, to the right hand side of Equation 3.9. The conservation of energy manifests itself in Equation 3.10. We recognize

$-P/\rho \vec{\nabla} \cdot \vec{v}$ as the specific rate of doing PdV work. Λ represents the cooling function that includes heating and cooling contributions. In the case of an isothermal approximation we neglect this term. The ideal gas law (Equation 3.12) serves as the closure equation. For more information on the equation of state see Section 3.4.

3.3.2 Jeans Criterion

A thorough investigation of the stability of a homogeneous spherical density fluctuation of radius r requires a linear stability analysis. For the case of a non-magnetic, isothermal, infinite, homogeneous, self-gravitating medium at rest (i.e. without turbulent motions) Jeans (1902) derived a relation between the oscillation frequency ω and the wave number k of small perturbations,

$$\omega^2 - c_s^2 k^2 + 4\pi G \rho_0 = 0, \qquad (3.13)$$

where c_s is the isothermal sound speed, G the gravitational constant, and ρ_0 the initial mass density. The derivation neglects viscous effects and assumes that the linearized version of the Poisson equation describes only the relation between the perturbed potential and the perturbed density (neglecting the potential of the homogeneous solution, the so-called 'Jeans swindle'; see, for example, Binney & Tremaine, 1987). The third term in Equation 3.13 is responsible for the existence of decaying and growing modes, as pure sound waves stem from the dispersion relation $\omega^2 - c_s^2 k^2 = 0$. Perturbations are unstable against gravitational contraction if their wave number is below a critical value, the Jeans wave number k_J, i.e. if

$$k^2 < k_J^2 \equiv \frac{4\pi G \rho_0}{c_s^2}, \qquad (3.14)$$

or equivalently, if the wavelength of the perturbation exceeds a critical size given by $\lambda \equiv 2\pi k_J^{-1}$. Assuming the perturbation is spherical with diameter λ_J, this directly translates into a mass limit

$$M_J \equiv \frac{4\pi}{3} \rho_0 \left(\frac{\lambda_J}{2} \right)^3 = \frac{\pi}{6} (\frac{\pi}{G})^{3/2} \rho_0^{-1/2} c_s^3. \qquad (3.15)$$

All perturbations exceeding the Jeans mass M_J will collapse under their own weight. For isothermal gas $c_s^2 \propto T$, so $M_J \propto \rho_0^{-1/2} T^{3/2}$. The critical mass M_J decreases when the density ρ_0 grows or when the temperature T sinks.

3.3.3 Turbulence

Turbulence is nonlinear fluid motion resulting in velocity modes over a wide range of spatial and temporal scales. Turbulence is deterministic and unpredictable, but it is not reducible to a low-dimensional system and so does not exhibit the properties of classical chaotic dynamical systems. The strong correlations and lack of scale separation preclude the truncation of statistical equations at any order.

Hydrodynamic turbulence arises because the nonlinear advection operator, $(\vec{v} \cdot \vec{\nabla})\vec{v}$ (see Equation 3.9), generates severe distortions of the velocity field by stretching, folding, and dilating fluid elements. The effect can be viewed as a continuous set of topological deformations of the velocity field (Ottino, 1989), but in a much higher dimensional space than chaotic systems, so that the velocity field is, in effect, a stochastic field of nonlinear straining. These distortions self-interact to generate large amplitude structure covering the available range of scales. For incompressible turbulence (i.e. constant density) driven at large scales, this range is called the inertial range (see below) because the advection term corresponds to inertia in the equation of motion. For a purely hydrodynamic incompressible system, this range is measured by the ratio of the advection term to the viscous term, which is the Reynolds number

$$Re = \frac{vL}{\nu}, \quad (3.16)$$

where v and L are the characteristic large-scale velocity and length and ν is the kinematic viscosity. In the cool ISM, $Re \approx 10^5$ to 10^7 if viscosity is the damping mechanism. With compressibility, magnetic fields, or self-gravity, all the associated fields are distorted by the velocity field. Hence, one can have MHD turbulence, gravitational turbulence, or thermally driven turbulence, but they are all fundamentally tied to the advection operator. These additional effects lead to fundamental changes in the behavior of the turbulence. This may affect the way energy is distributed among scales, which is often referred to as the turbulent cascade. For a more detailed discussion of the complex statistical characteristics of turbulence, we refer the reader to the book by Lesieur (1997).

Most studies of turbulence treat incompressible turbulence, characteristic of most terrestrial applications. Root-mean-square (rms) velocities are subsonic, and the density remains almost constant. Dissipation of energy occurs primarily in the smallest vortices[2], where the dynamical scale L becomes comparable to the length on which viscosity acts $\mathcal{L}_{\text{visc}}$. Kolmogorov (1941) described a heuristic theory based on dimensional analysis that captures the basic behavior of incompressible turbulence surprisingly well, although subsequent work has refined the details substantially. He assumed turbulence driven on a large scale L, forming eddies at that scale. These eddies interact to form slightly smaller eddies, transferring some of their energy to the smaller scale. The smaller eddies in turn form even smaller ones, until energy has cascaded all the way down to the dissipation scale $\mathcal{L}_{\text{visc}}$. In order to maintain a steady state, equal amounts of energy must be transferred from each scale in the cascade to the next, and eventually dissipated, at a rate

$$\dot{E} = \frac{\eta v^3}{L} \quad (3.17)$$

where η is a constant determined empirically. This leads to a power-law distribution of kinetic energy $E \propto v^2 \propto k^{-11/3}$, where $k = 2\pi/L$ is the wave number, and density does not enter because of the assumption of incompressibility. Most of the energy remains near the driving scale, while energy drops off steeply below $\mathcal{L}_{\text{visc}}$. Because of the apparently local nature of the cascade in wave number space, the viscosity only determines the behavior of the energy distribution at the bottom of the cascade on scales below $\mathcal{L}_{\text{visc}}$, while the driving only determines the behavior near the top of the cascade for spatial scales at and above L.

---

[2] A vortex can be any circular or rotational flow that possesses non-zero vorticity. Vorticity $\vec{\omega}$ is defined as $\vec{\omega} = \vec{\nabla} \times \vec{v}$, where $\vec{v}$ is the fluid velocity.

The region in between is known as the inertial range, in which energy transfers from one scale to the next without influence from driving or viscosity. The behavior of the flow in the inertial range can be studied regardless of the actual scale at which L and $\mathcal{L}_{\text{visc}}$ lie, so long as they are well separated.

3.3.4 Interstellar Turbulence

Gas flows in the ISM vary, however, from the above idealized picture in three important ways. First, they are highly compressible, with Mach numbers $\mathcal{M}$ ranging from order unity in the warm (10^4 K), diffuse ISM, up to as high as 50 in cold (10 K), dense molecular clouds. Second, the equation of state of the gas is very soft due to radiative cooling, so that pressure $P \propto \rho^\gamma$ with the polytropic index falling in the range $0.4 < \gamma < 1.2$ as a function of density and temperature (see, for example, Scalo et al., 1998; Ballesteros-Paredes et al., 1999b; Spaans & Silk, 2000). Third, the driving of the turbulence is not uniform, but rather comes from inhomogeneous processes, e.g. supernovae (e.g. Norman & Ferrara, 1996; Mac Low, 1999).

Supersonic flows in highly compressible gas create strong density perturbations. Early attempts to understand turbulence in the ISM (von Weizsäcker, 1943, 1951; Chandrasekhar, 1949) were based on insights drawn from incompressible turbulence. An attempt to analytically derive the density spectrum and resulting gravitational collapse criterion was first made by Chandrasekhar (1951b,a). This work was followed up by several authors, culminating in work by Sasao (1973) on density fluctuations in self-gravitating media whose interest has only recently been appreciated. Larson (1981) qualitatively applied the basic idea of density fluctuations driven by supersonic turbulence to the problem of star formation. Bonazzola et al. (1992) used a renormalization-group technique to examine how the slope of the turbulent velocity spectrum could influence gravitational collapse. This approach was combined with low-resolution numerical models to derive an effective adiabatic index for subsonic compressible turbulence by Panis & Pérault (1998). Adding to the complexity of the problem, the strong density inhomogeneities observed in the ISM can be caused not only by compressible turbulence, but also by thermal phase transitions (Field et al., 1969; McKee & Ostriker, 1977; Wolfire et al., 1995) or gravitational collapse (Kim & Ostriker, 2001).

In supersonic turbulence, shock waves offer additional possibilities for dissipation. Shock waves can also transfer energy between widely separated scales, removing the local nature of the turbulent cascade typical of incompressible turbulence. The energy spectrum may change only slightly, however, as the Fourier transform of a step function representative of a perfect shock wave is k^{-2}. Integrating in three dimensions over an ensemble of shocks, one finds the differential energy spectrum $E(k)dk = \rho v^2(k)k^2 dk \propto k^{-2} dk$. This is just the compressible energy spectrum reported by Porter & Woodward (1994); Porter et al. (1992, 1994). They also found that even in supersonic turbulence, the shock waves do not dissipate all the energy, as rotational motions continue to contain a substantial fraction of the kinetic energy, which is then dissipated in small vortices. Boldyrev (2002) has proposed a theory of velocity structure function scaling based on the work of She & Leveque (1994) using the assumption that dissipation in supersonic turbulence primarily occurs in sheetlike shocks, rather than linear filaments at the centers of vortex tubes. The first comparisons to numerical models show good agreement with this model (Boldyrev et al., 2002a), and it has been extended to the density structure functions by Boldyrev et al. (2002b). Transport properties of supersonic turbulent flows in the astrophysical context have been discussed by de Avillez & Mac Low (2002) and Klessen & Lin (2003). The driving of interstellar turbulence is neither uniform nor homogeneous. Controversy still reigns over the most important energy sources at different scales. For a more detailed discussion of ISM turbulence see Elmegreen & Scalo (2004) and Scalo & Elmegreen (2004).

3.4 Equation of State

As mentioned in Section 3.3.1, the set of hydrodynamical equations has to be closed by an equation of state (EOS), which describes the relation between pressure P, volume V and temperature T of the gas. Typical for a closure equation, it cannot be derived from within the system, but stems from including additional physical phenomena connected to the properties of the matter. In the simplest description, we consider the ISM as an ideal gas:

$$P = \mathcal{R}\rho T(\rho) = \frac{k_\mathrm{B}}{\mu m_\mathrm{p}} \rho T(\rho), \tag{3.18}$$

where ρ is the density. The parameters k_B, μ, m_p are Boltzmann constant, molecular weight and proton mass. The ideal gas law can also be written as

$$P = (\gamma_\mathrm{ad} - 1)u\rho, \tag{3.19}$$

where γ_ad is the ratio of specific heats[3] of a substance at constant pressure and constant volume. The variable u represents the specific internal energy (energy per unit mass). This approximation is roughly accurate for any classical system, composed of particles not interacting at a molecular level at low pressure and high temperature. In the following subsections we discuss two astrophysically relevant approximations to the EOS.

3.4.1 Isothermal Equation of State

For the densities and temperatures in molecular clouds, i.e. $1\,\mathrm{cm}^{-1} \leq n(\mathrm{H}_2) \leq 10^7\,\mathrm{cm}^{-3}$ and $T \approx 10\,\mathrm{K}$, the gas can cool very efficiently and the opacities in the molecular lines involved are low enough for the medium to be optically thin. Hence, treating the gas isothermally is a good first approximation (e.g. Hayashi & Nakano, 1965; Hayashi, 1966; Larson, 1969, 1973b, see also the discussion in Section 2.2). With the isothermal sound speed $c_\mathrm{s} = (\mathcal{R}T)^{1/2}$ it follows that

$$P = c_\mathrm{s}^2 \rho \text{ and } \gamma = 1 \tag{3.20}$$

3.4.2 Polytropic Equation of State

The true nature of the EOS remains a major theoretical problem in understanding the fragmentation properties of molecular clouds. Some calculations invoke cooling during the collapse (Monaghan & Lattanzio, 1991; Turner et al., 1995; Whitworth et al., 1995). Others include radiation transport to account for the heating that occurs once the cloud reaches densities of $n(\mathrm{H}_2) \geq 10^{10}\,\mathrm{cm}^{-3}$ (Myhill & Kaula, 1992; Boss, 1993), or simply assume an adiabatic equation of state once this density is exceeded (Bonnell, 1994; Bate et al., 1995). Spaans & Silk (2000) showed that radiatively cooling gas can be described by a polytropic

---
[3]The specific heat (also called specific heat capacity) is the amount of heat required to change a unit mass (or unit quantity, such as mole) of a substance by one degree in temperature.

EOS, in which the polytropic exponent γ changes with gas density ρ. Considering a polytropic EOS is still a rather crude approximation. In practice the behavior of γ may be more complicated and important effects like the temperature of the dust, line-trapping and feedback from newly-formed stars should also be taken into account (Scalo et al., 1998). Nevertheless a polytropic EOS gives an insight into the differences that a departure from isothermality evokes.

Following the considerations in Section 2.2, we use a polytropic equation of state to describe the thermal state of the gas in our models (see Chapter 6) with a polytropic exponent that changes at a certain critical density ρ_c from γ_1 to γ_2

$$P = K_1 \rho^{\gamma_1} \quad \rho \leq \rho_c$$
$$P = K_2 \rho^{\gamma_2} \quad \rho > \rho_c, \quad (3.21)$$

where K_1 and K_2 are constants, and P and ρ are the thermal pressure and gas density. For an ideal gas, the equation of state is

$$P = \frac{k_B}{\mu m_p} \rho T(\rho), \quad (3.22)$$

where T is the temperature, and k_B, μ, and m_p are the Boltzmann constant, molecular weight, and proton mass. So the constants K_1 and K_2 can be written as

$$K_1 = \frac{k_B}{\mu m_p} \rho^{1-\gamma_1} T_1(\rho) K_2 = \frac{k_B}{\mu m_p} \rho^{1-\gamma_2} T_2(\rho). \quad (3.23)$$

Since K_1 and K_2 are defined as constants in ρ, it follows for T

$$T_1 = a_1 \rho^{\gamma_1 - 1} \quad \rho \leq \rho_c$$
$$T_2 = a_2 \rho^{\gamma_2 - 1} \quad \rho > \rho_c, \quad (3.24)$$

where a_1 and a_2 are constants. The initial conditions define a_1

$$a_1 = T_0 \rho_0^{1-\gamma_1} \quad (3.25)$$

At ρ_c it holds that

$$T_1(\rho_c) = T_2(\rho_c) \quad (3.26)$$

Thus, a_2 can be written in terms of a_1

$$a_2 = a_1 \rho_c^{\gamma_1 - \gamma_2}. \quad (3.27)$$

According to the analytical work by Jeans (1902) on the stability of a self-gravitating, isothermal medium the oscillation frequency ω and the wave number k of small perturbations satisfy the dispersion relation in Equation 3.13. The perturbation is unstable if the wavelength λ exceeds the Jeans length $\lambda_J = 2\pi/k_J$ or, equivalently, if the mass exceeds the Jeans mass in Equation 3.15. In a system with a polytropic EOS, i.e. $P = K\rho^\gamma$, the sound speed is

$$c_s = \left(\frac{dP}{d\rho}\right)^{1/2} = (K\gamma)^{1/2} \rho^{(\gamma-1)/2}. \quad (3.28)$$

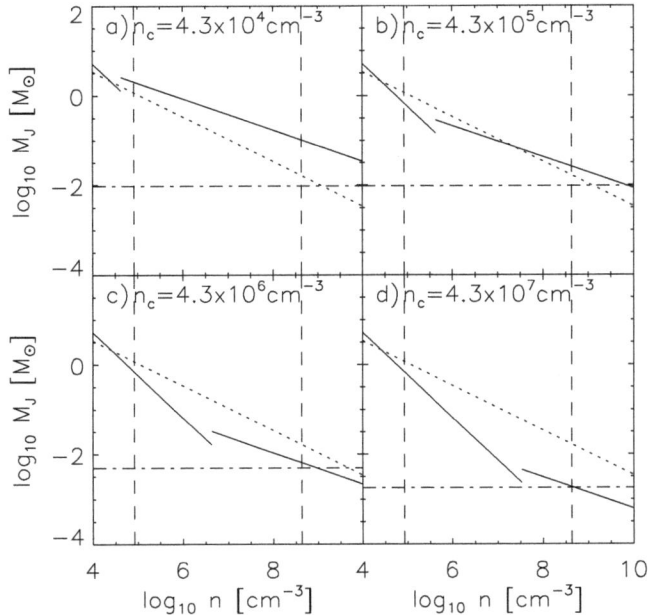

Figure 3.1: Local Jeans mass as a function of density for four different critical densities n_c. For comparison the dependence is also shown for the isothermal case (dotted line). The Jeans mass changes at the critical density. The initial mean density and the density at which sink particles form are represented by the vertical dashed lines. The dashed-dotted lines show the minimal resolvable mass for the runs with the highest resolution. For a description of the simulations see Chapter 6. From Jappsen et al. (2005).

Thus, the Jeans mass can be written as

$$M_\text{J} = \frac{\pi^{5/2}}{6} \left(\frac{K}{G}\right)^{3/2} \gamma^{3/2} \rho^{(3/2)\gamma - 2} \ . \tag{3.29}$$

Using equations 3.23, 3.24, 3.25, 3.27 and 3.29 one finds

$$M_\text{J1} = \frac{\pi^{5/2}}{6} \left(\frac{k_\text{B} T_0 \rho_0^{1-\gamma_1}}{G \mu m_\text{p}}\right)^{3/2} \gamma_1^{3/2} \rho^{(3/2)\gamma_1 - 2} \qquad \rho \leq \rho_\text{c},$$

$$M_\text{J2} = \frac{\pi^{5/2}}{6} \left(\frac{k_\text{B} T_0 \rho_0^{1-\gamma_1}}{G \mu m_\text{p}}\right)^{3/2} \gamma_2^{3/2} \rho_\text{c}^{(3/2)(\gamma_1 - \gamma_2)} \rho^{(3/2)\gamma_2 - 2} \qquad \rho > \rho_\text{c}.$$

The sound speed changes when the polytropic index changes at ρ_c, so M_J also varies such that

$$\frac{M_\text{J1}}{M_\text{J2}} = \left(\frac{\gamma_1}{\gamma_2}\right)^{\frac{3}{2}} . \tag{3.30}$$

In Figure 3.1 we see this change as a small jump in the evolution of M_J with density. This jump occurs at the critical density where we change from polytropic index γ_1 to polytropic index γ_2. If we use $\gamma_1=0.7$ and $\gamma_2=1.1$ as justified in Section 2.2 then $M_{J1} \propto \rho^{-0.95}$.

During the initial phase of collapse, the turbulent flow produces strong ram pressure gradients that form density enhancements. Higher density leads to smaller local Jeans masses, so these regions begin to collapse and fragment. Simulations with an SPH code different from the one used in the present work show that fragmentation occurs more efficiently for smaller constant values of γ, and less efficiently for $\gamma > 1$, cutting off entirely at $\gamma > 1.4$ (Li et al., 2003; Arcoragi et al., 1991). For filamentary systems, fragmentation already stops for $\gamma > 1$ (Kawachi & Hanawa, 1998). For a more thorough discussion see Chapter 6.

3.5 Timescales

At this point we summarize some important timescales for future reference. Comparing the timescales enables us to determine the physical processes that influence the dynamical evolution of star-forming gas.

- Free-fall Time:
  A homogeneous sphere of pressureless material of density ρ_0 will collapse to a point within a free-fall time t_{ff}, where

$$t_{\text{ff}} = \sqrt{\frac{3\pi}{32G\rho_0}}. \tag{3.31}$$

  For a derivation see Stahler & Palla (2004).

- Sound-crossing Time:
  The time a molecular cloud needs to react to a change in pressure is given by the sound-crossing time t_s, which is approximately given by the time a sound wave needs to cross the cloud of size L

$$t_s = \frac{L}{c_s} = L\left(\frac{dP}{d\rho}\right)^{-1/2}. \tag{3.32}$$

  For an isothermal situation with a sound speed $c_s = \sqrt{\mathcal{R}T}$ this simplifies to

$$t_s = \frac{L}{\sqrt{\mathcal{R}T}}. \tag{3.33}$$

- Turbulent Crossing Time:
  The time t_{tc} in which matter can be transported by a turbulent flow. It is given by

$$t_{\text{tc}} = \frac{L}{v_{\text{rms}}} = \frac{L}{\mathcal{M}c_s}, \tag{3.34}$$

where L, $v_{\rm rms}$ and $\mathcal{M}$ represent, respectively, the size of the cloud, the root mean square velocity of the gas and the associated Mach number. Thus $t_{\rm tc}$ and $t_{\rm s}$ are connected via

$$\frac{t_{\rm s}}{t_{\rm tc}} = \mathcal{M}. \tag{3.35}$$

- Cooling Time:
  For our simulations with an implicit calculated cooling function (see Chapter 7) we also have to take into account the time it takes a cloud of a certain temperature T to cool to zero temperature if there are no heat sources:

$$t_{\rm cool} \equiv \frac{E_{\rm int}}{|\dot{E}_{\rm cool}|}. \tag{3.36}$$

For a monatomic gas with a total number density of particles $n_{\rm tot}$, and for Λ being the cooling rate in units of $\rm erg\,cm^{-3}\,s^{-1}$, the cooling time is given by

$$t_{\rm cool} \simeq \frac{\frac{3}{2} n_{\rm tot} kT}{\Lambda}. \tag{3.37}$$

The cooling function Λ depends on the temperature T and the number density $n_{\rm tot}$.

- Hubble Time:
  For any cosmological simulation the Hubble time is of importance. The distance between two fundamental observers $l(t)$ changes with time t according to

$$l(t) = l(t_0) R(t), \tag{3.38}$$

where t_0 denotes the present time and $R(t)$ is the scale factor, which is a universal function with $R(t_0) = 1$. The expansion rate $H(t)$ is defined by

$$H(t) \equiv \frac{\dot{R}(t)}{R(t)}. \tag{3.39}$$

At the present time, $H(t_0) = \dot{R}(t_0) \equiv H_0$ is the Hubble constant. The Hubble time $t_{\rm H}$ can be expressed at any redshift z as

$$t_{\rm H} = \frac{1}{H(t)} = \frac{1}{H_0 (1+z)^{3/2}}. \tag{3.40}$$

Chapter 4

The Numerical Scheme

Most of the calculations presented in this work were done with the publicly available code gadget by Springel et al. (2001) which we have modified as explained in this chapter. gadget evolves collisionless particles with the traditional N-body approach, and self-gravitating gas by smoothed particle hydrodynamics (SPH), as explained in Section 4.1. gadget features a parallel version that has been designed to run on massively parallel supercomputers with distributed memory. It uses a tree algorithm[1] (octal-tree, for details see Barnes & Hut, 1986) to compute gravitational forces. Periodic boundary conditions are supported by means of an Ewald summation technique (Hernquist et al., 1991). The code uses individual and adaptive timesteps for all particles, and it combines this with a scheme for dynamic tree updates. Due to its Lagrangian nature, gadget thus allows a very large dynamic range to be bridged, both in space and time. For further detail see the code paper by Springel et al. (2001). Because we are interested in gravoturbulent fragmentation, we include turbulence in our version of the code that is driven uniformly with the method described by Mac Low et al. (1998) and Mac Low (1999). We give more details in Section 3.3.3. During gravoturbulent fragmentation it is necessary to follow the gas over several orders of magnitude in density. SPH simulations of collapsing regions become slower as more particles move to higher density regions and hence have small timesteps. Replacing dense cores by accreting sink particles leads to considerable increase of the overall computational performance. In Section 4.3 we introduce the concept of sink particles which we have implemented into gadget and which allows us to follow the dynamical evolution of the system over many free-fall times. In Section 4.5 we also give an overview of the chemistry and cooling routines that we have added to the code.

4.1 Smoothed Particle Hydrodynamics

Smoothed Particle Hydrodynamics (SPH) (Gingold & Monaghan, 1977; Lucy, 1977) is a computational method used for simulating fluid flows. It has been used in many fields

---

[1] In tree methods the particles are arranged in a hierarchy of groups. When the force on a particular particle is computed, the force exerted by distant groups is approximated by their lowest multipole moments.

of research, including astrophysics, ballistics, vulcanology and tsunami research. It is a Lagrangian method and the resolution of the method can easily be adjusted with respect to variables such as the density.

The SPH method works by dividing the fluid into a set of discrete 'fluid elements'. These particles have a spatial distance (known as the 'smoothing length', typically represented in equations by h), over which their properties are 'smoothed' by a kernel function. Any physical quantity at any position $\vec{r}$ can be obtained by summing the relevant properties of all the particles which lie within a smoothing length[2]. For example, the temperature of particle i depends on the temperatures of all the particles within a radial distance h of particle i. The contributions of each particle to a fluid property are weighted according to their distance from the particle of interest. Mathematically, this is given by the kernel function W which is usually 1-dimensional, i.e. assumes spherical symmetry. There are a variety of appropriate functions proposed in the literature, ranging from Gaussian functions (Gingold & Monaghan, 1977), to spline functions of third or higher order and with compact support (e.g. Monaghan & Lattanzio, 1985; Monaghan, 1985). These kernels interpolate at least to second order in h, and are always positive in the range of interest. Furthermore, all are smooth functions with well defined first derivatives. The spline functions have the advantage, that there is a clear limit to the number of particles contributing to the average process due to their compact support. For others one has to implement an artificial cut-off. In gadget the smoothing kernel is a spline of the form (Monaghan & Lattanzio, 1985):

$$W(r,h) \equiv \frac{8}{\pi h^3} \begin{cases} 1 - 6\left(\frac{r}{h}\right)^2 + 6\left(\frac{r}{h}\right)^3 & \text{for } 0 \leq \frac{r}{h} \leq \frac{1}{2} \\ 2\left(1 - \frac{r}{h}\right)^3 & \text{for } \frac{1}{2} < \frac{r}{h} \leq 1 \\ 0 & \text{for } \frac{r}{h} > 1 \end{cases} \quad (4.1)$$

Note that the smoothing kernel in gadget is defined in the interval $[0, h]$ and not on $[0, 2h]$ as it is frequently done in other SPH calculations.

In SPH, the value of any quantity A at a position $\vec{r}$ is given by the equation

$$A(\vec{r}) = \sum_i m_i \frac{A_i}{\rho_i} h_i^{-3} W\left(\frac{|\vec{r} - \vec{r}_i|}{h_i}\right), \quad (4.2)$$

where m_i is the mass of particle i, A_i is the value of the quantity A for particle i, ρ_i is the density associated with particle i, and W is the kernel function mentioned above. For example, the density at position $\vec{r}$ can be expressed as:

$$\rho(\vec{r}) = \sum_i m_i h_i^{-3} W\left(\frac{|\vec{r} - \vec{r}_i|}{h_i}\right), \quad (4.3)$$

where the summation over i includes all particles in the simulation. Similarly, the spatial derivative of a quantity can be obtained by using integration by parts to shift the ∇-operator from the physical quantity to the kernel function,

$$\nabla A(\vec{r}) = \sum_i m_i \frac{A_i}{\rho_i} h_i^{-4} W'\left(\frac{|\vec{r} - \vec{r}_i|}{h_i}\right) \frac{\vec{r} - \vec{r}_i}{|\vec{r} - \vec{r}_i|}, \quad (4.4)$$

---

[2]The number of smoothing lengths used is just convention and can change between different implementations.

where $W'(s) \equiv \mathrm{d}W/\mathrm{d}s$. Nevertheless in practice we apply a spline kernel with compact support (see definition of the kernel in Equation 4.1). The summation becomes finite due to the fact that the kernel function has compact support. This implies that we take only contributions from a few close neighboring particles into account. It has proven useful to determine h such that the number of neighbors always lies in the range from 30 to 70. In our simulations it varies between 35 and 45. The fact that the summation does not need to be over all particles greatly reduces the computational costs of all SPH calculations. Although the size of the smoothing length can be fixed in both space and time, this does not take advantage of the full power of SPH. By assigning each particle its own smoothing length and allowing it to vary with time, the resolution of a simulation can be made to automatically adapt itself depending on local conditions. For example, in a very dense region where many particles are close together the smoothing length can be made relatively short, yielding high spatial resolution. Conversely, in low-density regions individual particles are far apart and the resolution is low, optimizing the computation for the regions of interest.

However, variable smoothing lengths require special care to ensure force antisymmetry. When applying the smoothing procedure in the most straightforward way, the mutual forces between two particles i and j are no longer necessarily anti-symmetric for different smoothing lengths h_i and h_j. For example, particle i will "see" particle j, but the reverse need not be the case. Newton's third law is violated and momentum is no longer a conserved quantity. Following Benz (1990), gadget solves this problem by simply replacing h in all previous equations by the arithmetic average of the smoothing length for all particle pairs,

$$h \longrightarrow h_{ij} = \frac{h_i + h_j}{2}. \qquad (4.5)$$

Combined with an equation of state and an appropriate time integration scheme, SPH can simulate hydrodynamic flows efficiently. However, SPH tends to smear out shocks and contact discontinuities to a much greater extent than state-of-the-art grid-based schemes. The Lagrangian-based adaptivity of SPH is analogous to the adaptivity present in grid-based adaptive mesh refinement codes, though in the latter case one can refine the mesh spacing according to any criterion selected by the user. Because it is Lagrangian in nature, SPH is limited to refining based on density alone.

Often in astrophysics, one wishes to model self-gravity in addition to pure hydrodynamics. The particle-based nature of SPH makes it ideal to combine with a particle-based gravity solver, for instance tree gravity, particle mesh, or particle-particle mesh.

4.2 Resolution Issues

All numerical simulations, Eulerian or Lagrangian, have to be able to prevent the non-physical growth of numerical perturbations and to resolve formed structures. Only then can they give reliable and realistic results. What does his mean for SPH simulations?

SPH particles are allowed to move around and increase the resolution where needed (e.g. regions of high density). The artificial growth of perturbations is inhibited, provided that the

SPH calculations give accurate estimates for these properties, i.e. there is adequate sampling of the fluid with enough particles of similar properties within each kernel as described in the preceding section. Bate & Burkert (1997) demonstrated that adequate sampling means in this case that the local Jeans mass M_J should be resolved at all times. By this they mean that the minimum resolvable mass by SPH, $M_{\rm res}$ should fulfill the condition

$$2M_{\rm res} \lesssim M_J. \tag{4.6}$$

Hereby $M_{\rm res}$ denotes the mass within the radius of a kernel, with $M_{\rm res} = N_{\rm neigh} m_{\rm part}$, where $N_{\rm neigh}$ is the number of neighbors within a kernel (~ 40 in our simulations) and $m_{\rm part}$ the mass of each SPH gas particle. With this Jeans condition and the definition of the isothermal Jeans mass (Equation 3.15) we find the following relation between the total number of particles N in a simulation with total gas mass M and the maximum resolvable density $\rho_{\rm max}$:

$$2N_{\rm neigh}\frac{M}{N} \lesssim M_J = \frac{\pi}{6}\left(\frac{\pi}{G}\right)^{3/2}\rho_{\rm max}^{-1/2}c_s^3 \iff \rho_{\rm max} = \frac{\pi^5 c_s^6}{144 G^3 M^2 N_{\rm neigh}^2}N^2 \tag{4.7}$$

For a polytropic equation of state the Jeans mass changes for a polytropic index $\gamma < 4/3$ according to Equation 3.29 and therefore the number of particles needed rises accordingly for decreasing M_J. For example, Figure 3.1 shows the values for $M_{\rm res}$ (marked by the horizontal line) in our simulations discussed in Section 6.

4.3 Sink Particles

4.3.1 The Concept

Numerical simulations of star formation in a turbulent environment lead often to increasing densities in unpredictable regions of the computational volume. As we have seen in Section 4.2 there is a maximum resolvable density at which we have to stop the simulations to guarantee reliable results. Moreover the growing density contrast demands an ever decreasing minimum time step.

Introducing sink particles allows us to follow the dynamical evolution of the system over many free-fall times. We have specifically included sink particles in the gadget code. Once the density contrast in the center of a collapsing cloud core exceeds the maximum resolvable density $\rho_{\rm max}$, as calculated in the preceding section in Equation 4.7, the entire central region of the core is replaced by a 'sink particle' (Bate, Bonnell & Price, 1995). It is a single, non-gaseous, massive particle that only interacts with normal SPH particles via gravity. Gas particles that come within a certain radius of the sink particle, the accretion radius $r_{\rm acc}$, are accreted if they are bound to the sink particle. This allows us to keep track of the total mass, the linear and angular momenta of the collapsing gas.

Each sink particle defines a control volume with a fixed radius. This radius is chosen to be the Jeans length at the threshold density $\rho_{\rm max}$, following Bate & Burkert (1997). We cannot resolve the subsequent evolution in its interior. Combination with a detailed one-dimensional implicit radiation hydrodynamic method shows that a protostar forms in the very center about 10^3 yr after sink creation (Wuchterl & Klessen, 2001). We subsequently call the sink protostellar object or simply protostar. Altogether, the sink particle represents only the innermost, highest-density part of a larger collapsing region.

In reality, however, at some late phases of the collapse additional effects become important. Conservation of angular momentum enforces the formation of an accretion disk, where matter can only stream toward the center on a viscous time scale. Magnetic fields play an important role and may drive outflows along the spin axis. Finally at the very center, the density might reach the level at which nuclear fusion sets in. All these effects are not included in our code. We are interested in the earlier phases of fragmentation and collapse.

4.3.2 Sink Particles in a Parallel Code

The parallel version of gadget distributes the SPH particles onto the individual processors, using a spatial domain decomposition. Each processor hosts a rectangular piece of the computational volume. If the position of a sink particle is near the boundary of this volume, the accretion radius overlaps with domains on other processors. We therefore communicate the data of the sink to all processors. Each processor searches for gas particles within the accretion radius of the sink. Three criteria determine whether the particle gets accreted or not. First, the particle must be bound to the sink particle, i.e., the kinetic energy must be less than the magnitude of the gravitational energy. Second, the specific angular momentum of the particle must be less than what is required to move on a circular orbit with radius r_{acc} around the sink particle. Finally, the particle must be more tightly bound to the candidate sink particle than to other sink particles. If all tests are satisfied the gas particle is considered accreted, i.e. its mass, velocity and momenta are added to those of the sink particle.

Once the central region of a collapsing gas clump exceeds a given density contrast $\Delta\rho/\rho$, we introduce a new sink particle. The procedure for dynamically creating a sink particle is as follows. We search all processors for the gas particle with the highest density. When this density is above the threshold and when its smoothing length is less than half the accretion radius, then the gas particle is considered a candidate sink particle. If the accretion radius around the candidate particle overlaps with another domain, its position is sent to the other processors. Every processor searches for the particles that exist in its domain and, simultaneously, within the accretion radius of the candidate particle. These particles and the candidate particle undergo a series of tests to decide if they should form a sink particle. First, the new sink particle must be the only sink within two accretion radii. Second, the ratio of thermal energy to the magnitude of the gravitational energy must be less than 0.5. Third, we require that the total energy is less than zero. Finally, the divergence of the accelerations on the particles must be less than zero. If all these tests are passed, the particle with the highest density turns into a sink particle with position, velocity and acceleration derived from the center of mass values of the original gas particles within $r_{\rm acc}$. If these original particles are distributed over several processors the center of mass values have to be communicated correctly to the processor that hosts the new sink particle.

Ideally, the creation of sink particles in an SPH simulation should not affect the evolution of the gas outside its accretion radius. In practice there is a discontinuity in the SPH particle distribution due to the hole produced by the sink particles. This affects the pressure and viscous forces on particles outside the sink. We have implemented adequate boundary conditions at the 'surface' of the sink particles as described in detail in Bate et al. (1995) to correct for these effects.

Following Bate et al. (1995) we use the Boss & Bodenheimer (1979) standard isothermal test case for the collapse and fragmentation of an interstellar cloud core to check our sink implementation. Initially, the cloud core is spherically symmetric with a small $m = 2$ perturbation and uniformly rotating. As gravitational collapse proceeds a rotationally supported high-density bar builds up in the center embedded in a disk-like structure. The two ends of the bar become gravitationally unstable, resulting in the formation of a binary system. We see no further subfragmentation (see also, Truelove et al., 1997). These tests show that the precise creation time and the mass of the sink particle at the time of its formation can vary somewhat with the number of processors used. We also find that simulations with different processor numbers show small deviations in the exact positions and velocities of the gas particles. These variations are due to the differences in the extent of the domain on each processor. When the force on a particular particle is computed, the force exerted by distant groups is approximated by their lowest multipole moments. Since each processor constructs its own Barnes & Hut tree, differences in the tree walk result in differences in the computed force. Hence, the formation mass and time of sink particles depend on the computational setup. Nevertheless, these differences are only at the 0.1% level and the total number of collapsing objects is not influenced by changes in the number of processors.

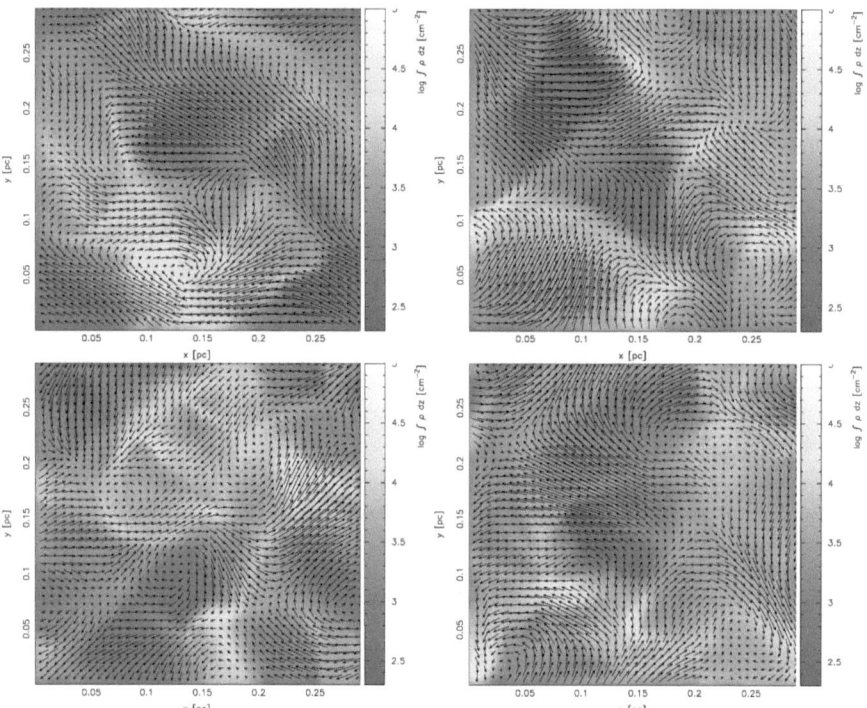

Figure 4.1: Column density and velocity field in the xy-plane of the runs R6k2, R6k2r1, R6k2r2 and R6k2r3 (for details on the simulations see Chapter 6 and Table 6.1). The 4 runs differ in the realizations of the Gaussian velocity field but they have the same Mach number and driving scale. All runs are plotted at the same time before self-gravity has been turned on, but at a time when dynamical equilibrium has been reached.

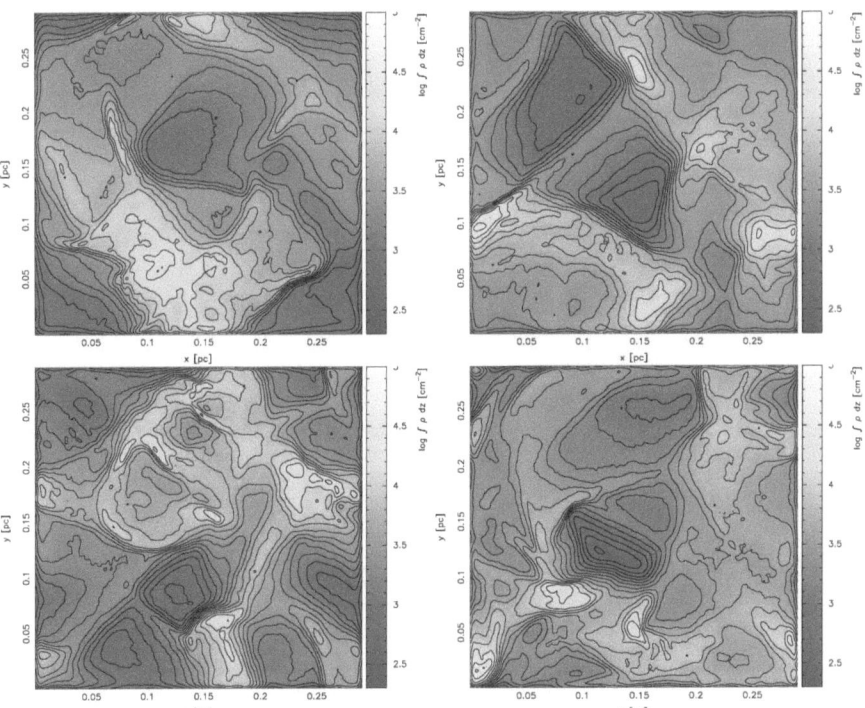

Figure 4.2: Column density and isodensity contours in the xy-plane for the same runs as depicted in Figure 4.1. All runs are plotted at the same time before self-gravity has been turned on, but at a time when dynamical equilibrium has been reached. This is to illustrate various effects (see Section 4.4).

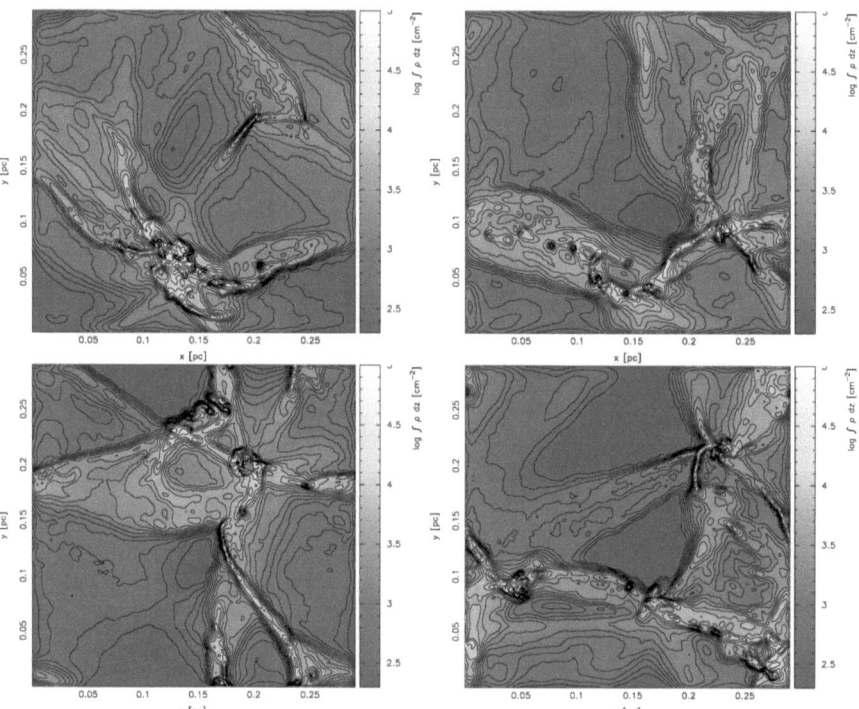

Figure 4.3: Column density and isodensity contours in the xy-plane for the same runs as depicted in Figure 4.1. All runs are plotted at the time where 50 % of the gas has been accreted onto sink particles.

4.4 Turbulent Driving

We generate turbulent flows by introducing Gaussian velocity fluctuations. We include this scheme specifically to the standard gadget code. We usually start our simulations with a homogenous density distribution and zero velocity. We then add velocities according to a random velocity field which we calculate separately for the three velocity components v_x, v_y, v_z. A Gaussian field $p(\vec{r})$ at position $\vec{r}$ is completely characterized by its first two moments, its mean value $p_0 \equiv \langle p(\vec{r}')\rangle_{\vec{r}'}$ and its standard deviation $\sigma(\vec{r}) \equiv \langle p(\vec{r}')p^*(\vec{r}'+\vec{r})\rangle_{\vec{r}'}$, which is equivalent to the Fourier transform of the power spectrum $P(\vec{k})$ (Lesieur, 1997). For an isotropic fluctuation spectrum we get $P(k) = P(|\vec{k}|)$. By defining a normalization p_0 and a power spectrum $P(k)$ in Fourier space, all statistical properties of the field $p(\vec{r})$ are determined.

The values $P(k)$ specify the contribution of wave numbers k to the statistical fluctuation spectrum. In Gaussian random fields, the phases are arbitrarily chosen from a uniform distribution in the interval $[0, 2\pi]$, and the amplitudes for each mode k are randomly drawn from a Gaussian distribution with width $P(k)$ centered on zero. Since waves are generated from random processes, the properties of an ensemble of fluctuation fields are determined only in a statistical sense. Individual realizations may deviate considerably from this mean value, especially at small wave numbers k, i.e. at long wavelengths, where only a few modes (k_x, k_y, k_z) contribute to a wave number $k = |\vec{k}|$.

In most of our simulations we use a power spectrum

$$P(k) \propto k^N, \tag{4.8}$$

with N=0. Additionally, we only drive on scales with wave numbers in a narrow interval $k-1 \leq |\vec{k}| \leq k$, where $k = L/\lambda_d$ counts the number of driving wavelengths λ_d in a box of size L (Mac Low et al., 1998). This offers a simple approximation to driving mechanisms that act on a single scale. To drive the turbulence, this fixed pattern is normalized to maintain constant kinetic-energy input rate $\dot{E} = \Delta E/\Delta t$ (Mac Low, 1999). Self-gravity is turned on only after the turbulence reaches a state of dynamical equilibrium. The mean root square velocity of the simulated gas has then reached a certain Mach number (see Figure 4.4).

As an example we show the velocity fields for 4 different runs in Figure 4.1. All models shown are driven with velocity fields that have the same power spectrum that acts on scales between $k = 1$ and $k = 2$ and are driven with the same energy rate. They differ only in the random number that was used to generate the Gaussian velocity field. Figure 4.2 shows the density distribution after dynamical equilibrium has been achieved, but without considering self-gravity of the gas. All runs show mainly large scale structures as is expected for this scale of driving. Density enhancements occur at different locations due to the different converging turbulent flows. Figure 4.3 shows the density distribution of the same runs but now with self-gravity included. At the time depicted the sink particles have accreted approximately 50 % of the gas. Filamentary structure is visible in all runs, but the spatial distribution differs. This also influences the time and the number of sink particle creation.

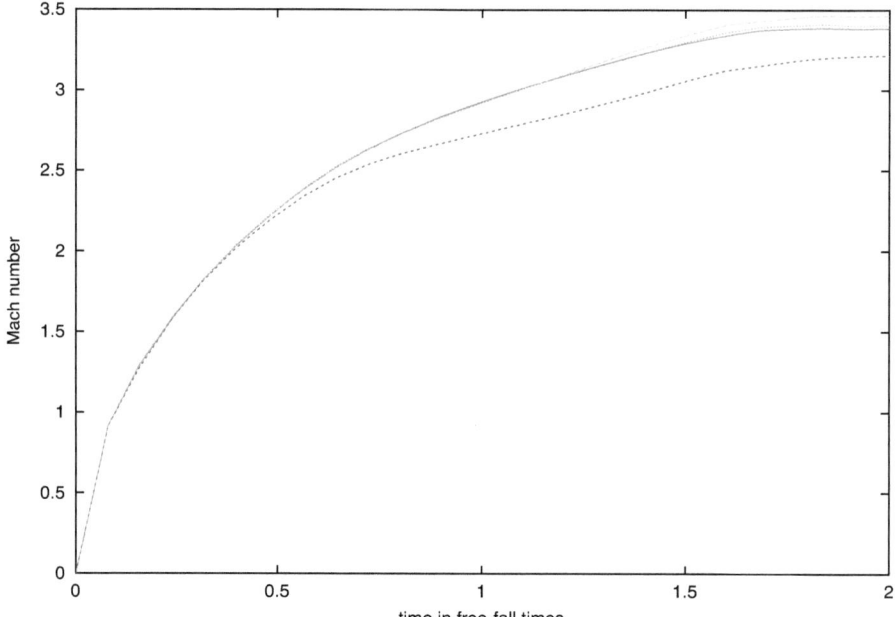

Figure 4.4: Root mean square velocity of the gas particles as a function of time from the same runs as shown in Figure 4.1 with different realizations of the Gaussian velocity field but with the same driving scale.

4.5 Chemistry and Cooling

In this section we give an overview of our additions to gadget that allow us to follow the dominating cooling processes and chemical reactions in protogalactic gas of zero or low metallicity (see Chapter 7). In large regions of our parameter space of density and temperature the cooling is dominated by the non-equilibrium chemistry of H_2 (Glover, 2003). To follow the chemistry, we associate a set of chemical abundances with each SPH particle. Just as with the other fluid properties, such as density or internal energy, these abundances represent averages over the local fluid flow (see Section 4.1). For each SPH particle, we therefore must solve a set of chemical equations of the form

$$\frac{\mathrm{d}x_i}{\mathrm{d}t} = C_i - D_i x_i, \qquad (4.9)$$

where x_i is the fractional abundance of species i, computed with respect to the total number density of hydrogen nuclei (i.e., $x_i = n_i/n$, where n_i is the number density of species i and n is the number density of hydrogen nuclei), and where C_i and D_i are terms representing the chemical creation and destruction of species i. The values of the creation and destruction terms generally depend on both the density and temperature of the gas, as well as on its chemical composition.

In our code we track the abundances of H_2, H^+, C ii, O ii and Si ii out of equilibrium. Using the conservation laws for charge and total number of nuclei of the elements, we also follow H, e^-, C, O and Si. Because H^- and H_2^+ reach equilibrium on very short time scales, we do not attempt to follow their equilibrium directly in our simulations. Instead, their abundances are computed only as required, under the assumption that both reach equilibrium instantaneously. In the case of H^-, one can readily show that equilibrium is reached on a timescale

$$t_{\rm eq, H^-} \simeq (k_{\rm ad} n_{\rm H} + k_{\rm mn} n_{\rm H^+})^{-1}, \tag{4.10}$$

where $n_{\rm H}$ is the hydrogen atom density, $n_{\rm H^+}$ is the H^+ density, and $k_{\rm ad}$ and $k_{\rm mn}$ are the associative detachment and mutual neutralization rate coefficients, respectively. We assume that other processes responsible for destroying H^-, such as photodetachment (reaction 6 in Table B.1) can be neglected. If $n_{\rm H^+} \ll (k_{\rm ad}/k_{\rm mn}) n_{\rm H}$ then the associative detachment term dominate and since $k_{\rm ad} \approx 10^{-9} \, {\rm cm}^3 {\rm s}^{-1}$ (to within an order of magnitude), it follows that $t_{\rm eq, H^-} \approx 10^9 n_{\rm H}^{-1}$ s. On the other hand, if $n_{\rm H^+} \gg (k_{\rm ad}/k_{\rm mn}) n_{\rm H}$, then the mutual neutralization term dominates, in which case $t_{\rm eq, H^-} \approx 5 \times 10^5 \sqrt{T} n_{\rm H^+}^{-1}$ s, again to within an order of magnitude. In either case, the equilibrium timescale is shorter than either the cooling time and the free-fall time of the gas (which are typically much longer than a Myr). In more ionized gas, or at higher densities, $t_{\rm eq, H^-}$ becomes even smaller relative to the cooling time $t_{\rm cool}$ and the free-fall time $t_{\rm ff}$.

The case of H_2^+ is very similar: one can show that it reaches chemical equilibrium on a timescale
$$t_{eq,H_2^+} \simeq (k_{ct} n_H + k_{dr} n_{e^-})^{-1}, \qquad (4.11)$$
where k_{ct} and k_{dr} are the rate coefficients of the charge transfer and dissociative recombination reactions which are the main processes responsible for removing H_2^+ from the gas. Evaluating this expression, we obtain $t_{eq,H_2^+} \approx 1.6 \times 10^9 n_H$ s in the low ionization limit, or $t_{eq,H_2^+} \approx 1.6 \times 10^8 n_{e^-}$ s in the high ionization limit. Again, these timescale are both very much shorter than either the cooling time or the free-fall time of the gas.

Because H^- and H_2^+ reach equilibrium so quickly, we do not attempt to follow their chemical equilibrium directly in our simulations. Instead, their abundances are computed only as required, under the assumption that both reach equilibrium instantaneously. This approximation introduces a certain amount of error into the computed H_2 abundance. However, provided that the timesteps used to evolve the SPH particles in the simulation are long compared to t_{eq,H^-} and t_{eq,H_2^+}, which we have verified is the case for our simulations, the size of this error is negligible compared to that arising from the rate coefficient uncertainties (Glover, 2005).

To solve the chemical equations for a given SPH particle, we make use of a technique known as operator splitting. We assume that within the current timestep of the SPH particle, we can solve for the density evolution of the gas separately from its chemical evolution. The density evolution can then be computed using the same prescription as in standard SPH (see for details Springel et al., 2001), and the updated gas density is available for use in the chemical equations. These are then solved implicitly using dvode, a freely available and well tested double precision ordinary differential equation solver (Brown et al., 1989). Operator splitting introduces a certain amount of error, as in practice the density should vary during the chemical timestep. However, the SPH algorithm naturally limits the extent to which the density can change during a single SPH particle timestep, by making particles take shorter timesteps in rapidly evolving regions. We therefore expect the error introduced by this technique to be small.

In common with other authors, we use a simplified reaction network that does not include the chemistry of minor coolants such as HD or LiH. We also neglect any effects due to the helium chemistry. Neglect of the minor coolants is justified by the fact that H_2 dominates the cooling of the gas for all of the temperatures and densities found in our simulations (Flower & Pineau des Forêts, 2001). At worst, we may overestimate the final temperature of the gas slightly.

Neglect of the helium chemistry is also easily justified, provided that we assume that the bulk of the helium in the gas is in neutral form, as in this case the only reactions involving helium that play any role in determining the H_2 abundance – the collisional dissociation of H^- and H_2 by He – are far less effective than the corresponding reactions with H (Abel et al., 1997). So the error in the final H_2 abundance will be small. If significant amounts of ionized helium are present, then our assumption introduces a larger error, since we will underestimate the actual electron abundance in the gas, and hence the H^- formation rate. However, even in this case, we would expect the error in the final H_2 abundance to be relatively small, owing to the small abundance of helium relative to hydrogen. The chemical reactions included in our network are summarized in Table B.1. In most cases, we also list the reference used for the adopted rate coefficient. The exceptions are the associative detachment and mutual neutralization reactions, which we discuss below.

Table B.1 lists three photochemical reactions: the photodetachment of H^- and the photodissociation of H_2^+ and H_2. When simulating protogalaxies (see Section 7), we assume illumination by an external background radiation spectrum with the shape of a 10^5 K blackbody, as should be typical of the brightest population III stars (Cojazzi et al., 2000). We cut off this spectrum at energies greater than 13.6 eV to account for absorption by neutral hydrogen in the protogalactic gas and in the intergalactic medium. The strength of the background is specified in terms of the flux at the Lyman limit,

$$J(\nu_\alpha) = 10^{21} J_{21} \mathrm{erg\, s^{-1}\, cm^{-2}\, Hz^{-1}\, sr^{-1}}. \tag{4.12}$$

If sufficient H_2 forms within the protogalaxy, it will begin to self-shield, reducing the effective photodissociation rate. An exact treatment of the effects of self-shielding is computationally unfeasible, as it would require us to solve for the full spatial, angular and frequency dependence of the radiation field at every timestep. Instead, we have chosen to incorporate it in an approximate manner. We assume that the dominant contribution to the self-shielding at a given point in the protogalaxy comes from gas close to that point, and so only include the contribution to the self-shielding that comes from the nearby H_2. To implement this approximation numerically, we make use of the fact that gadget already defines a suitable local length scale: the SPH smoothing length h, which characterizes the scale over which the flow variables are averaged. In gadget, as in all modern SPH codes, h is allowed to vary from point to point within the flow and is automatically adjusted in order to keep the mass enclosed within a sphere of radius h roughly constant (see Section 4.1). Further details can be found in Springel et al. (2001).

In our calculation of the H_2 column density used to compute the degree of self-shielding, we include only H_2 that lies within a single smoothing length of the point of interest. We justify this approximation by noting that in our simulations, widely separated SPH particles typically move with a significant velocity relative to one another. Consequently, the contribution to the total absorption arising from one particle is Doppler shifted when viewed from the rest frame of the other particle. If this Doppler shift is large compared to the line widths of the Lyman-Werner band transitions, then the effect will be to dramatically reduce the extent to which the absorption contributes to the total self-shielding. On the other hand, gas close to the point of interest will typically have a much smaller relative velocity, and so will contribute far more effectively. Our approximation considers only the latter contribution, and assumes that the former contribution is completely negligible. In practice, of course, the distant gas is likely to contribute to some non-negligible extent, and so we will tend to underestimate the true amount of self-shielding, unless the gas infall is highly supersonic. Nevertheless, we believe that our approximation remains useful as it is computationally efficient, and also represents an improvement over previous optically thin treatments (e.g. Machacek et al., 2001; Ricotti et al., 2002). Finally, we assume that ionization from X-rays or cosmic rays is negligible, although previous work suggests that even if a low level of ionization is present, it will not have a major effect on the outcome of the collapse (Glover & Brand, 2003; Machacek et al., 2003).

A second major modification that we have made to the gadget code is the inclusion of a treatment of radiative heating and cooling. Cooling in our model comes from three main sources: electron impact excitation of atomic hydrogen (a.k.a. Lyman-f cooling), which is effective only above about 8000 K, rotational and vibrational excitation of H_2, and Compton cooling. Rates for Lyman-f cooling and Compton cooling were taken from Cen (1992), while for H_2 ro-vibrational cooling we used a cooling function from Le Bourlot et al. (1999). In models where an ultraviolet background is present, we include the effects of heating from the photodissociation of H_2, assuming that 0.4 eV of energy per photodissociation is deposited as heat (Black & Dalgarno, 1977). We also include heating due to the ultraviolet pumping of H_2, following Burton et al. (1990), although this is only important in high density gas ($n \geq 10^4 \text{cm}^{-3}$). To incorporate the radiative heating and cooling terms into gadget, we again use an operator splitting technique. In this case, we assume that the change in the internal energy of the gas due to pressure work can be computed separately from that due to radiative heating and cooling. The former can then be calculated in the same fashion as in the standard gadget code, while the latter can be computed by solving:

$$\frac{d\epsilon}{dt} = \Gamma - \Lambda, \tag{4.13}$$

where ϵ is our initial estimate of the internal energy density of the gas, which already includes the effects of the pressure work term. Γ represents the heating rate per unit volume and Λ represents the cooling rate per unit volume. We solve this equation implicitly using dvode at the same time that we solve the chemical equations. Again, the use of operator splitting introduces some error into the thermal evolution of the gas, but, as before, we expect this error to be small. For a complete list of the chemical reactions and further details on the implementation we refer the reader to Appendix B.

4.6 Periodic Boundary Conditions

Each choice of boundary conditions possesses a number of advantages and disadvantages. In all our simulations we use fully periodic boundary conditions. The fully periodic choice constrains the distribution of matter in a cube of finite size to appear periodic and implies particular assumptions about long-range tidal effects. These observationally unjustified assumptions give rise to an anisotropy in the force field due to the fact that there are more particles along diagonal lines of the periodized cube than along the edges. However, our choice is motivated by the demands of the problems that we investigate. Simulating the dynamical evolution and fragmentation in the interior of a molecular cloud, a process that finally leads to the formation of new stars, requires periodic boundaries to prevent the whole object from collapsing to the center. The situation is quite similar in cosmological large-scale structure simulations, where periodic boundaries mimic the homogeneity and isotropy of the initial matter distribution. Hernquist et al. (1991) showed that in simulations with fully periodic boundary conditions the linear evolution of density fluctuations agrees well with analytic calculations. In Gadget periodic boundaries are implemented by means of the Ewald summation technique (Hernquist et al., 1991; Springel et al., 2001). This technique has been used extensively in solid-state physics, especially in molecular dynamics, where periodic boundaries are a natural requirement.

Chapter 5

Angular Momentum Evolution

In this chapter we investigate the rotational properties and angular momentum evolution of prestellar and protostellar cores formed from gravoturbulent fragmentation of interstellar gas clouds using hydrodynamic simulations. We find the specific angular momentum j of molecular cloud cores in the prestellar phase to be on average $\langle j \rangle = 7 \times 10^{20}\,\mathrm{cm^2\,s^{-1}}$ in our models. This is comparable to the observed values. A fraction of those cores are gravitationally unstable and go into collapse to build up protostars and protostellar systems, which then have $\langle j \rangle = 8 \times 10^{19}\,\mathrm{cm^2\,s^{-1}}$. This is one order of magnitude lower than that of their parental cores and in agreement with observations of main-sequence binaries. The loss of specific angular momentum during collapse is mostly due to gravitational torques exerted by the ambient turbulent flow as well as by mutual protostellar encounters in a dense cluster environment. Magnetic torques are not included in our models, these would lead to even larger angular momentum transport.

The ratio of rotational to gravitational energy β in cloud cores that go into gravitational collapse turns out to be similar to the observed values. We find that β is roughly conserved during the main collapse phase. This leads to the correlation $j \propto M^{2/3}$, between specific angular momentum j and core mass M. Although the temporal evolution of the angular momentum of individual protostars or protostellar systems is complex and highly time-variable, this correlation holds well in a statistical sense for a wide range of turbulent environmental parameters. In addition, high turbulent Mach numbers result in the formation of more numerous protostellar cores with, on average, lower mass. Therefore, models with larger Mach numbers result in cores with lower specific angular momentum. We find, however, no dependence of the angular momentum on the spatial scale of the turbulence. Our models predict a close correlation between the angular momentum vectors of neighboring protostars during their initial accretion phase. Possible observational signatures are aligned disks and parallel outflows. The latter are indeed observed in some low-mass isolated Bok globules.

5.1 Initial Conditions and Model Parameters

The suite of models considered here consists of 12 numerical simulations where turbulence is maintained with constant root-mean-square Mach numbers in the range $2 \leq \mathcal{M} \leq 10$. This roughly covers the range observed in typical Galactic molecular clouds. All our simulations start from a cube with homogeneous density and periodic boundary conditions. We apply a non-local scheme that inserts energy in a limited range of wave numbers at a given rate (see Section 4.4).We distinguish between turbulence that carries its energy mostly on large scales, at wave numbers $1 \leq k \leq 2$, on intermediate scales, i.e. $3 \leq k \leq 4$, and on small scales with $7 \leq k \leq 8$. The corresponding wavelengths are $\ell = L/k$, where L is the total size of the computational volume. The models are labeled mnemonically as M$\mathcal{M}$kk, with rms Mach number $\mathcal{M}$ and wave number k. After 1.5 free-fall times we turn on self-gravity and continue driving. Each of these simulations contains 205379 SPH particles. We also consider a model that is globally unstable and contracts from an initial Gaussian random density field without turbulence (for details see Klessen & Burkert, 2000, 2001). It is called GA and was run with 500000 particles. The main parameters are summarized in Table 5.1. Note that the final star formation efficiency varies between the different models, as indicated in column 5 of Table 5.1. This simply reflects the evolutionary stage at the time when we stop the calculation. In some cases the accretion timescale is too long to follow the simulation to high efficiencies.

Table 5.1: Sample parameters, name of the environment used in the text consisting of the Mach number $\mathcal{M}$ and the driving scale k (GA denotes the model with initial Gaussian density), number $\mathcal{N}$ of protostellar objects (i.e. sink particles in the centers of protostellar cores) at the final stage of the simulation, percentage of accreted mass at the final stage M_{acc}/M_{tot}, parameter $\mathcal{A}$ see Equation 5.5, parameter $\mathcal{B}$ see Equation 5.9

Name	k	$\mathcal{M}$	$\mathcal{N}$	M_{acc}/M_{tot} [%]	$\mathcal{A}$	$\mathcal{B}$
					[10^{20} cm^2 s^{-1}]	
M2.0k2	1..2	2.0	68	75	1.7	1.7
M2.0k4	3..4	2.0	62	48	2.0	2.0
M2.0k8	7..8	2.0	11	66	1.7	1.6
M3.2k2	1..2	3.2	62	80	1.2	1.3
M3.2k4	3..4	3.2	37	82	2.0	1.9
M3.2k8	7..8	3.2	17	60	2.7	2.6
M6k2	1..2	6.0	110	76	1.3	1.3
M6k4	3..4	6.0	60	65	1.5	1.7
M6k8	7..8	6.0	7	4	1.9	1.4
M10k2	1..2	10.0	100	38	1.0	1.0
M10k4	3..4	10.0	10	6	2.0	1.4
M10k8	7..8	10.0	27	8	1.4	1.05
GA	...	...	56	85	1.4	1.05

5.1.1 Physical scaling and naming convention

In this chapter we use a self-gravitating, isothermal model that studies the interplay between gravity and gas pressure, it is therefore scale free. Besides the initial conditions, the dynamical evolution of the system depends only on one parameter, namely the ratio between internal and potential energy. This ratio can be interpreted as dimensionless temperature $\mathcal{T} \equiv \epsilon_{\text{int}}/|\epsilon_{\text{pot}}|$. Line widths in molecular clouds are super-thermal, implying the presence of supersonic turbulent motions (Section 2.1). In case of isotropic turbulence, these non-thermal (turbulent) contributions can be accounted for by $\mathcal{T}$, i.e. introducing a second parameter which can be absorbed by defining an effective temperature $\mathcal{T} = T + \sigma_{\text{turb}}^2/\mathcal{R}$, where $\mathcal{R}$ is the universal gas constant. The turbulent velocity dispersion is denoted σ_{turb}. In case of anisotropic turbulent motions, the system has (locally) preferred axes and the concept of one single effective temperature is no longer valid.

Scaled to physical units we adopt a gas temperature of $11.3\,\text{K}$ corresponding to a sound speed $c_s = 0.2\,\text{km}\,\text{s}^{-1}$, and we use a mean density of $n(\text{H}_2) = 10^5\,\text{cm}^{-3}$, which is typical for star-forming molecular cloud regions (e.g. in ρ-Ophiuchi, see Motte, André & Neri, 1998). The mean thermal Jeans mass (see Section 3.3.2) in all models is $\langle M_{\text{J}} \rangle = 1\,M_\odot$. The turbulent models contain a mass of $120\,M_\odot$ within a cube of size $0.29\,\text{pc}$, and the Gaussian model has $220\,M_\odot$ in a volume of $(0.34\,\text{pc})^3$. The global free-fall timescale is $\tau_{\text{ff}} = 10^5\,\text{yr}$, and the simulations cover a density range from $n(\text{H}_2) \approx 100\,\text{cm}^{-3}$ in the lowest density regions to $n(\text{H}_2) \approx 10^9\,\text{cm}^{-3}$ where the central parts of collapsing gas clumps are converted into sink particles as described in Section 4.3.

Each sink particle defines a control volume with a fixed radius of $560\,\text{AU}$. We cannot resolve the subsequent evolution in its interior. After $\sim 10^3\,\text{yr}$ a protostar will form in the very center of the sink. Because of angular momentum conservation most of the matter that falls in will assemble in a protostellar disk. It is then transported inward by viscous and possibly gravitational torques (e.g. Bodenheimer, 1995; Papaloizou & Lin, 1995; Lin & Papaloizou, 1996). With typical disk sizes of the order of several hundred AU, the control volume fully encloses both star and disk. If low angular momentum material is accreted, the disk is stable and most of the material ends up in the central star. In this case, the disk simply acts as a buffer and smooths possible accretion spikes. It will not delay or prevent the mass growth of the central star by much. However, if material that falls into the control volume carries large specific angular momentum, then the mass load onto the disk is likely to be faster than the inward transport. The disk grows large and may become gravitationally unstable and fragment. This will lead to the formation of a binary or higher-order multiple (Bodenheimer et al., 2000).

Throughout the next sections, we adopt the following naming convention: In the pre-collapse phase, we call high-density gas clumps prestellar cores or simply gas clumps. They build up at the stagnation points of converging flows. The flows result from turbulent motion that establishes a complex network of interacting shocks. The gas clumps are identified and characterized using a three-dimensional clump-finding algorithm as described in Appendix A of Klessen & Burkert (2000). The density fluctuations in turbulent velocity fields are highly transient. They can disperse again once the converging flow fades away. Even clumps that are strongly dominated by gravity may get disrupted by the passage of a new shock front.

For local collapse to result in the formation of stars, Jeans-unstable, shock-generated density fluctuations must therefore collapse to sufficiently high densities on time-scales shorter than the typical time-interval between two successive shock passages. We include in our analysis of the angular momentum only Jeans-unstable gas clumps. Angular momentum is calculated from the internal clump motions with respect to the location of the density maximum. These objects correspond to the so called starless cores observed e.g. by Goodman et al. (1993), Barranco & Goodman (1998), Jijina et al. (1999), and others. They are thought to collapse and build up a central protostar or protostellar system in the later stages of evolution. Once collapse has led to the formation of an embedded protostar (in our scheme identified by a central sink particle) we call the object protostellar core or also protostar. The angular momentum is obtained as the spin accumulated by the sink particle during its accretion history. The angular momentum distribution is best compared with observations of main-sequence binaries as we expect the unresolved star-disk system interior to the sink particle to break up into a binary or higher-order multiple (see Section 5.3).

5.2 Molecular Cloud Clumps and Prestellar Cores

Figure 5.1 shows the distribution of the specific angular momenta of the gas clumps that were identified in the turbulent environment M6k2 (see Table 5.1). We compare with observational values of prestellar cores taken from Caselli et al. (2002) for (a), from Goodman et al. (1993), Barranco & Goodman (1998), Jijina et al. (1999) for (b) and from Pirogov et al. (2003) for (c). According to Goodman et al. (1993) the values for the specific angular momenta are obtained by using best-fit velocity gradients from maps of observed line-center velocities under the assumption of solid body rotation. The cores in Goodman et al. (1993), Barranco & Goodman (1998) and Jijina et al. (1999) were mapped in the (J,K) = (1,1) transition of NH_3, whereas the massive cloud cores in Pirogov et al. (2003) and the dense cloud cores in Caselli et al. (2002) were mapped in N_2H^+. As found in Caselli et al. (2002), the two molecular species trace essentially the same material, especially in starless cores. Model clumps and observed molecular cloud clumps have comparable mass spectra (Klessen, 2001) and similar shapes (Klessen & Burkert, 2000; Ballesteros-Paredes et al., 2003). Figure 5.1 demonstrates that something similar holds for the distribution of specific angular momenta.

Note that, when transforming from dimensionless code units into physical scales, the specific angular momentum depends on the mean number density n and the temperature T as $j \propto T/\sqrt{n(H_2)}$. In Figure 5.1a, we use our standard scaling (see Section 5.1) corresponding to regions like the ρ Ophiuchi main cloud (Motte et al., 1998). This is adequate for the low-mass cores studied by Caselli et al. (2002), and in Section 5.3 we will furthermore show that the specific angular momenta of collapsed cores fall into the right range for main-sequence binaries.

We find that the specific angular momenta of prestellar cores have values between $1 \times 10^{20}\,\text{cm}^2\,\text{s}^{-1}$ and $5 \times 10^{21}\,\text{cm}^2\,\text{s}^{-1}$ with a mean value of approximately $\langle j \rangle \approx 5 \times 10^{20}\,\text{cm}^2\,\text{s}^{-1}$. This is in good agreement with the Caselli et al. (2002) sample which has $\langle j \rangle = 7 \times 10^{20}\,\text{cm}^2\,\text{s}^{-1}$. Their cloud cores have a mean mass of $\sim 6\,M_\odot$, comparable to the core masses in our simulations. A Kolmogorov-Smirnov (KS) test was performed. We find that at the 50 % level the distributions are statistically indistinguishable.

However, the cores observed by Goodman et al. (1993) and Pirogov et al. (2003) trace lower densities and have higher mean masses of around $50\,M_\odot$ and $500\,M_\odot$, respectively. In Figure 5.1b we therefore use $n(\text{H}_2) = 1 \times 10^4\,\text{cm}^{-3}$ and $T = 10\,\text{K}$ for the scaling of our dimensionless code units leading to $\langle j \rangle = 1.5 \times 10^{21}\,\text{cm}^2\,\text{s}^{-1}$. This matches the observations, since Goodman et al. (1993) find $\langle j \rangle = 6 \times 10^{21}\,\text{cm}^2\,\text{s}^{-1}$ and a median value of $3 \times 10^{21}\,\text{cm}^2\,\text{s}^{-1}$. The massive cores mapped by Pirogov et al. (2003) have higher velocity dispersions, higher kinetic temperatures $(20 - 50\,\text{K})$ and densities $n(\text{H}_2) \geq 1 \times 10^4\,\text{cm}^{-3}$. The resulting mean specific angular momentum is $\langle j \rangle = 1.5 \times 10^{22}\,\text{cm}^2\,\text{s}^{-1}$. Again, adequate scaling in Figure 5.1c results in higher values of j in our models and leads to better agreement with the observed distribution.

Given the simplified assumptions in our numerical models, we find remarkably good agreement with the observed specific angular momenta in the prestellar phase. Similar findings are reported by Gammie et al. (2003). Similar to our study, they follow the dynamical evolution of an isothermal, self-gravitating, compressible, turbulent ideal gas. However, they include the effects of magnetic fields and solve the equations of motion using a grid-based method (the well-known ZEUS code). Their approach is thus complementary to ours. The j distribution that results from their simulations peaks at $4 \times 10^{22}\,\text{cm}^2\,\text{s}^{-1}$. They fix the mean number density at $n(\text{H}_2) \approx 1.0 \times 10^2\,\text{cm}^{-3}$ and use $T = 10\,\text{K}$. Using the same physical scaling we get very similar values, i.e. $\langle j \rangle = 4 \times 10^{22}\,\text{cm}^2\,\text{s}^{-1}$.

This mean value of j also falls in the range of specific angular momenta of cores that form in simulations by Li et al. (2004). They also use a version of the ZEUS code (ZEUS-MP) to perform high-resolution, three-dimensional, super-Alfvénic turbulent simulations to investigate the role of magnetic fields in self-gravitating core formation within turbulent molecular clouds. Adopting the same physical scaling as in Gammie et al. (2003), the specific angular momentum of their cores takes values between $5 \times 10^{21}\,\text{cm}^2\,\text{s}^{-1}$ and $8 \times 10^{22}\,\text{cm}^2\,\text{s}^{-1}$.

5.3 Protostars and Protostellar Systems

Figure 5.2 shows the distribution of specific angular momenta of collapsed cores at four different stages of mass accretion, ranging from the stage at which 15 % of the total available mass in the molecular cloud has been accreted onto collapsed cores in Figure 5.2a to 60 % in Figure 5.2d. As in the preceding section we use values from model M6k2. While the distribution of j narrows during the collapse sequence, the mean specific angular momentum remains essentially at the same value $j = (8 \pm 2) \times 10^{19}\,\text{cm}^2\,\text{s}^{-1}$ with a range from $10^{18}\,\text{cm}^2\,\text{s}^{-1}$ to $5 \times 10^{20}\,\text{cm}^2\,\text{s}^{-1}$. The specific angular momenta of the protostellar cores in the considered model are approximately one order of magnitude smaller than the ones of the

Jeans-unstable clumps for the same model, but both distributions join without a gap. In a statistical sense, there is a continuous transition as loss of angular momentum occurs during contraction. The range of specific angular momenta of the protostellar cores agrees well with the observed values for binaries (e.g. Bodenheimer, 1995). For this reason we compare in Figures 5.2a–d the model distributions with observations of binaries among G-dwarf stars by Duquennoy & Mayor (1991) in Figure 5.2e and with observations of young binaries in the Taurus star forming region by Simon (1992) in Figure 5.2f.

Duquennoy & Mayor (1991) derived a Gaussian-type period P distribution for their sample. Based on this distribution we calculated the distribution of the specific angular momenta using the following equation (see also Kroupa, 1995b, Equation 10):

$$j = 6.23 \times 10^{18} \left(1 - e^2\right)^{1/2} P^{1/3} \frac{m_1 m_2}{(m_1 + m_2)^{4/3}} \left(\text{cm}^2\,\text{s}^{-1}\right) \qquad (5.1)$$

where masses are in $M_\odot$ and P in days. We used a primary mass m_1 of $1\,M_\odot$, a mean mass ratio between primary and secondary $q = m_2/m_1 = 0.25$ and a mean eccentricity $e = 0.31$.

The resulting distribution has a mean specific angular momentum of $1.6 \times 10^{20}\,\text{cm}^2\,\text{s}^{-1}$, which agrees well with the values from the protostars in the simulations. This also holds for Figure 5.2f which was taken from Figure 5 in Simon (1992). The mean specific angular momentum here has a value of $1.6 \times 10^{20}\,\text{cm}^2\,\text{s}^{-1}$. The agreement of the distributions was confirmed by a χ^2 statistical test. Since our numerical resolution is not sufficient to follow the subfragmentation of collapsing cores into binary or higher-order multiple systems, the time evolution of j is an important tool to evaluate our models. We see a clear progression from the rotational properties of gas clumps (as discussed in Section 5.2) to those of the resulting collapsed cores. A similar correlation is observed between the cloud cores (Caselli et al., 2002) and typical main-sequence binaries (Duquennoy & Mayor, 1991). The former may be the direct progenitors of the latter.

5.3.1 Example of the Angular Momentum Evolution of a Protostellar Core in a Cluster

The evolution of the specific angular momenta of individual protostellar cores can be very complex depending on the rotational properties of their environment. There is a strong connection to the time evolution of the mass accretion rate. In Figure 5.3 we select five collapsed cores in model M6k2 with about the same final mass. All of them show a similar evolution of the specific angular momentum with increasing mass (Figure 5.3a) and time (Figure 5.3b). Nevertheless there are visible differences in the details.

Initially, the specific angular momentum increases with growing mass. However, at later stages the evolution strongly depends on the secular properties of the surrounding flow. In cores 43, 96, and 101, for example, j decreases again after reaching a peak value, while for cores 17 and 47 j stays close to the maximum value. Depending on the specific angular momentum of the accreted material, the resulting protostellar disks are expected to evolve quite differently. For example, preliminary 2-dimensional hydrodynamic calculations show that core 17 and core 47 will probably develop a stable disk (Bodenheimer 2003, priv. comm.). Boss (1999) showed that a high value of the ratio of rotational to gravitational energy $\beta > 0.01$ leads to fragmentation whereas $\beta < 0.01$ results in a stable disk. The ratio β for the peak value of j is $\beta = 0.005$ for core 17 and $\beta = 0.003$ for core 43 (for the definition of β see Equation 5.10). On the other hand, core 101 will probably fragment into a binary star. It has a $\beta = 0.016$. Also, the disk of core 96 is highly unstable with corresponding $\beta = 0.016$. The evolution of core 43 has not yet been followed sufficiently long to determine whether it will fragment to form a binary star or not. These results show the importance of the specific angular momentum on the evolution of the protostellar object. A similar result was found by Boss (1999) for slowly rotating, magnetic clouds.

Figure 5.3b shows that the change in specific angular momentum is closely linked to the mass accretion. At the point in time where the mass accretion rate (dotted line) has a pronounced peak, the change in specific angular momentum is also significant. A high mass accretion rate can result in an increase of specific angular momentum. But as seen for core 96 in Figure 5.3b, a high mass accretion rate can also lead to a reduction of specific angular momentum. The exact evolution of the specific angular momentum is thus closely linked to the flow properties of the surrounding material.

5.3.2 Statistical Correlation between Specific Angular Momentum and Mass

As mentioned in the introduction to this chapter, we find a correlation between mass M and specific angular momentum j in a statistical sense. The result is depicted in Figure 5.4. It shows the angular momentum evolution as a function of mass for all 110 collapsed cores in model M6k2. In Figure 5.4a, following Goodman et al. (1993), we adopt rigid body rotation with constant angular velocity Ω and uniform core density ρ. With these assumptions the specific angular momentum j can be written as

$$j = p\Omega R^2. \tag{5.2}$$

For a uniform density sphere $p = \frac{2}{5}$. The mass M of a sphere with constant density ρ_0 is related to the radius R via

$$M = \frac{4\pi}{3}\rho_0 R^3. \tag{5.3}$$

From Equations 5.2 and 5.3 follows that j can be expressed as:

$$j = p\Omega \left(\frac{3}{4\pi\rho_0}\right)^{2/3} M^{2/3}. \tag{5.4}$$

Therefore we fit the average angular momentum with a function of the form

$$j = \mathcal{A}(M/M_\odot)^{2/3}, \tag{5.5}$$

where $\mathcal{A} = p\Omega \left(\frac{3M_\odot}{4\pi\rho_0}\right)^{2/3}$. The constant $\mathcal{A}$ in Figure 5.4a has a value of $(1.3 \pm 0.6) \times 10^{20}$ cm^2 s^{-1}. This fit formula can be applied to different turbulent cloud environments, and we list the corresponding values of $\mathcal{A}$ for our model suite in Table 5.1. Using the fitted value and the density $\rho_0 = 4 \times 10^{-15}\,\mathrm{g\,cm^{-3}}$ where protostellar cores are identified, we calculate an angular velocity $\Omega = 1.33 \times 10^{-11}\,\mathrm{s^{-1}}$.

In this picture the ratio of rotational to gravitational energy β can be written as

$$\beta = \frac{(1/2)I\Omega^2}{qGM^2/R} = \frac{3p}{8\pi q}\frac{\Omega^2}{\rho_0 G}, \tag{5.6}$$

where the moment of inertia is given by $I = pMR^2$, and $q = \frac{3}{5}$ is defined such that qGM^2/R represents the gravitational potential of a uniform density sphere. With the assumptions of constant angular velocity and uniform density it follows that β is also constant. With Ω and ρ_0 as above, we get values $\beta \approx 0.05$. Goodman et al. (1993) as well as Burkert & Bodenheimer (2000) derived scaling relations where β is independent of radius. Similar values for β were also found by Goodman et al. (1993) for the observed cloud cores. In good agreement with our calculations they found that all values are below 0.18 with the majority being under 0.05.

The fit in Figure 5.4a rests on the assumption of the collapse of an initially homogeneous sphere with constant angular velocity. Using sink particles, however, which have a constant radius, makes it necessary to examine another possibility. In Figure 5.4b we thus assume a constant radius R and a constant β. Choosing a constant β is supported by the observations as discussed above and by our simulations as we show below. With β and Equation 5.6 (which still holds) it follows that the angular velocity depends on the density as

$$\Omega = \sqrt{\frac{8\pi q}{3p}G\rho\beta}\,. \tag{5.7}$$

Thus, the angular velocity Ω is no longer a constant. This implies for the specific angular momentum j

$$j = p\Omega R^2 = \sqrt{2pRqG\beta}\sqrt{M}. \tag{5.8}$$

Following Equation 5.8 we fit our data in Figure 5.4b with a square root function

$$j = \mathcal{B}\sqrt{M/M_\odot}\,, \tag{5.9}$$

where the moment of inertia is given by $I = pMR^2$, and $q = \frac{3}{5}$ is defined such that qGM^2/R represents the gravitational potential of a uniform density sphere. With the above density and angular velocity we find from Equation 5.6 that $\beta \approx 0.05$. For the new fit we find a scaling factor $\mathcal{B} = (1.3 \pm 0.6) \times 10^{20}\,\mathrm{cm^2\,s^{-1}}$ in the mass range $0 \leq M/M_\odot \leq 1.7$. Again, this fit formula can be applied to different turbulent cloud environments, and we list the corresponding values of $\mathcal{B}$ for our model suite in Table 5.1.

The question remains if our simulations support the assumption of a constant β. In Figure 5.5 we compare the distribution of β measured by Goodman et al. (1993) with values we extract from our model. For the prestellar cores we use the definition

$$\beta = \frac{E_\mathrm{rot}}{E_\mathrm{grav}}, \tag{5.10}$$

and calculate rotational and potential energy, E_{rot} and E_{grav}, consistently from the full three-dimensional gas distribution of each gas clump. We do not adopt any assumption about symmetry and shape of the density and velocity structure. If we assume that each clump is spherical and has roughly constant density, as implied for example by Equation 5.6, then β is overestimated by a factor of 2.7. This shows the importance of taking the full three-dimensional clump structure into account when analyzing the rotational properties of molecular cloud cores. Both prestellar cores (i.e Jeans-unstable gas clumps) and protostellar cores in our model typically have $\beta < 0.3$ with similar distributions. Thus, β stays mainly in the interval $[0, 0.3]$ and in this sense it remains approximately constant during the collapse.

It should be noted in passing that we also looked at the density structure of purely hydrodynamic turbulence, i.e. without self-gravity. If we again perform a clump-decomposition of the density structure and compute hypothetical β-values, we find $\beta \approx 2$. This is indicative of the high degree of vorticity inherent in all turbulent flows. However, it also suggests that dense molecular cloud cores are strongly influenced by self-gravity. The fact that all cores in the observational sample have $\beta < 0.2$ implies that gravitational contraction is needed to achieve density contrasts high enough for sufficiently low β. This agrees with the picture of gravoturbulent fragmentation where molecular cloud structure as whole is dominated by supersonic turbulence but stars can only form in those regions where gravity overwhelms all other forms of support.

Comparing the two fits in Figure 5.4 shows that our first set of assumptions is a better representation of the data. This is especially true during the early accretion phase where we have good statistics. It applies to different turbulent cloud environments as well. We conclude that – in a statistical sense – the angular momentum evolution of collapsing cloud cores can be approximately described as contraction of initially constant-density spheres undergoing rigid body rotation with constant angular velocity. This is consistent with the fact that cores from gravoturbulent fragmentation follow a Bonnor-Ebert-type radial density profile (Ballesteros-Paredes et al., 2003) and have roughly constant density in their innermost regions. It also supports the assumptions adopted by Goodman et al. (1993) and Burkert & Bodenheimer (2000).

5.3.3 Dependence of the Specific Angular Momentum on the Environment

When we compare the results of our complete suite of numerical models (see Table 5.1) we find as a general trend that the average angular momentum falls with increasing Mach number. This is illustrated in Figure 5.6a. However, this follows mainly from the positive dependence of angular momentum on mass and from a correlation between average mass of the cloud cores and Mach number. As seen in Section 5.3.2, the specific angular momentum increases on average as the mass of the core rises. In environments with a low Mach number the mass growth of the cores is undisturbed over longer periods of time and so larger masses can accumulate. This can be inferred from Table 5.1, where we list both the number of cores and the accreted mass. For higher Mach numbers higher number of cores form with on average less mass. Thus, the average angular momentum is expected to decrease with increasing Mach number (Figure 5.6a).

To detect a direct dependence of the specific angular momentum on the Mach number we select cores that belong in a certain mass bin and average the specific angular momentum only over those cores. The results are shown in Figures 5.6b-5.6d. We find that independently of the Mach number there is in general little spread of specific angular momentum for different driving scales and different times in the accretion history . Low mass protostars (Figure 5.6b) are an exception. In low Mach number environments they show especially low angular momentum in the early accretion phase. However, within the error bars we do not find a further dependence of j on the Mach number.

Compared to the Gaussian collapse case, turbulent driving with small Mach numbers results in higher specific angular momenta. This is due to input of turbulent energy that can be converted into rotational energy if the turbulent velocities are not too high.

5.4 Loss of Angular Momentum during Collapse

We find the specific angular momentum j of molecular cloud cores in the prestellar phase to be on average $\langle j \rangle = 7 \times 10^{20}\,\mathrm{cm}^2\,\mathrm{s}^{-1}$ in our models. This is comparable to the observed values. A fraction of those cores is gravitationally unstable and goes into collapse to build up protostars and protostellar systems, which then have $\langle j \rangle = 8 \times 10^{19}\,\mathrm{cm}^2\,\mathrm{s}^{-1}$. This is one order of magnitude lower than their parental cores and in agreement with observations of main-sequence binaries. The loss of specific angular momentum during collapse is mostly due to gravitational torques exerted by the ambient turbulent flow as well as by mutual protostellar encounters in a dense cluster environment. In a semiempirical analysis of isolated binary star formation, Fisher (2004) presented the effects of turbulence in the initial state of the gas on the binary orbital parameters. These properties were in agreement with observations if a significant loss of angular momentum was assumed. Magnetic torques are not included in our models, these would lead to even larger angular momentum transport.

5.5 Orientation of Angular Momentum Vector

We find in our simulations that neighboring protostellar cores have similarly oriented angular momenta. In Figure 5.7 the correlation of specific angular momenta of different protostellar cores with respect to their orientations is shown as a function of distance. We calculate the scalar product $\vec{j} \cdot \vec{j}$ for all cores and average it over cores with similar distances between each other.

Figure 5.8a shows the spatial configuration in model M6k2 after 15 % of the available material has been accreted onto the protostellar cores. These cores form in small aggregates with diameters below 0.07 pc. The corresponding Figure 5.7a shows a spatial correlation of the specific angular momenta for small distances. The correlation length, defined as the maximum distance between cores at which the cores show a positive correlation in Figure 5.7a, is approximately 0.05 pc. Thus, correlation length and cloud size are closely connected. This can be understood in the following way. Within one molecular cloud clump neighboring cores accrete from the same reservoir of gas and consequently gain similar specific angular momentum. In the early phase of accretion we therefore expect disks and protostellar outflows of neighboring protostars to be closely aligned. Indeed, several examples of parallel disks and outflows have been reported in low-mass, isolated Bok globules by Froebrich & Scholz (2003), Kamazaki et al. (2003), Nisini et al. (2001), and Saito et al. (1995). Alternative explanations for the alignment of the symmetry axes of young stars include density gradients in the prestellar phase or the presence of strong magnetic fields. However, Ménard & Duchêne (2004) found that the disks of T-Tauri stars driving jets or outflows are perpendicular to the magnetic field but disks of T-Tauri stars without jet are parallel to the field lines. This is very puzzling, showing the complexity of the situation that will naturally arise in strongly turbulent flows.

During subsequent accretion the correlation length decreases to values below 0.015 pc (Figure 5.7b). This means that only close systems remain correlated (see also Figure 5.8b). This has three reasons. First, small N systems of embedded cores are likely to dissolve quickly as close encounters lead to ejection (e.g. Reipurth & Clarke, 2001). Only close binaries are able to survive for a long time (e.g. Kroupa, 1995a,b). The correlation length therefore decreases with time. Second, the same turbulent flow that generated a collapsing high-density clump in the first place may also disrupt it again before its gas is fully accreted. If the clump contains several protostars they will disperse, again decreasing the correlation. Third, the opposite may happen. Turbulence may bring in fresh gas. The protostars are then able to continue accretion, but the specific angular momentum of the new matter is likely to be quite different from the original material. As protostars accrete at different rates, we expect a spread in $\vec{j}$ to build up and the alignment may disappear. At later stages of the evolution, we expect that the correlation between the specific angular momenta of close protostellar objects disappear almost completely. This is evident in Figure 5.7c. Furthermore, Figure 5.8c demonstrates that most of the initial subclustering has disappeared by this late stage.

5.6 Summary and Conclusions

We studied the rotational properties and time evolution of the specific angular momentum of prestellar and protostellar cores formed from gravoturbulent fragmentation in numerical models of supersonically turbulent, self-gravitating molecular clouds. We considered rms Mach numbers ranging from 2 to 10, and turbulence that is driven on small, intermediate, and large scales, as well as one model of collapse from Gaussian density fluctuations without any turbulence. Our sample thus covers a wide range of properties observed in Galactic star-forming regions, however, our main focus lies in typical low- to intermediate-mass star-forming regions like ρ-Ophiuchi or Taurus.

With the appropriate physical scaling, we find the specific angular momentum j of prestellar cores in our models, i.e. cloud cores as yet without central protostar, to be on average $\langle j \rangle = 7 \times 10^{20}\,\mathrm{cm^2\,s^{-1}}$. This agrees remarkably well with observations of cloud cores by Caselli et al. (2002) or Goodman et al. (1993). Some prestellar cores go into collapse to build up stars and stellar systems. The resulting protostellar objects have on average $\langle j \rangle = 8 \times 10^{19}\,\mathrm{cm^2\,s^{-1}}$. This is one order of magnitude less, and falls into the range observed in G-dwarf binaries (Duquennoy & Mayor, 1991). Collapse induced by gravoturbulent fragmentation is accompanied by a substantial loss of specific angular momentum. This is mostly due to gravitational torques exerted by the ambient turbulent flow and due to close encounters occurring when the protostars are embedded in dense clusters. This eases the angular momentum problem in star formation as described in Section 2.4 without invoking the presence of strong magnetic fields.

The time evolution of j is intimately connected to the mass accretion history of a protostellar core. As interstellar turbulence and mutual interaction in dense clusters are highly stochastic processes, the mass growth of individual protostars is unpredictable and can be very complex. In addition, a collapsing cloud core can fragment further into a binary or higher-order multiple or evolve into a protostar with a stable accretion disk. It is the ratio of rotational to gravitational energy β that determines which route the object will take. This is seen in the turbulent cloud cores studied here as well as in simulations of isolated cores where magnetic fields are important (e.g. Boss, 1999). The β-distribution resulting from gravoturbulent cloud fragmentation reported here agrees well with β-values derived from observations (Goodman et al., 1993). The average value is $\beta \approx 0.05$. Note, that we find that the distribution of β stays essentially the same during collapse and accretion (see also Burkert & Bodenheimer, 2000; Goodman et al., 1993).

Although the accretion history and thus the evolution of the specific angular momentum of a single protostellar object is complex, we find a clear correlation between j and mass M. This can be interpreted conveniently assuming collapse of an initially uniform density sphere in solid body rotation. Our models of gravoturbulent cloud fragmentation are best represented by the relation $j \propto M^{2/3}$.

When prestellar cores form by compression as part of supersonically turbulent flows and then go into collapse and possibly break apart into several fragments due to the continuing perturbation by their turbulent environment, we expect neighboring protostars to have similarly oriented angular momentum, at least during their early phases of accretion. Star clusters form hierarchically structured, with several young stellar objects being embedded in the same clump of molecular cloud material. These protostars accrete from one common reservoir of gas and consequently gain similar specific angular momentum. Their disks and protostellar outflows, therefore, will closely align. Indeed, there are several examples of parallel disks and outflows seen in low-mass, isolated Bok globules (Froebrich & Scholz, 2003; Kamazaki et al., 2003; Nisini et al., 2001; Saito et al., 1995). During later phases of cluster formation, the initial substructure becomes erased by dynamical effects and the correlation between the angular momenta of neighboring protostars vanishes. This is in agreement with our numerical calculations of gravoturbulent cloud fragmentation. These show small groups of close protostellar objects that have almost aligned specific angular momenta at birth. As expected, the alignment occurs during the early phase of accretion as neighboring protostars accrete material from the same region with similar angular momentum. During the subsequent evolution the above correlation decreases. This is either because protostellar aggregates disperse, or because infalling new material with different angular momentum becomes distributed unevenly among the protostars.

Altogether, the process of gravoturbulent fragmentation, i.e. the interplay between supersonic turbulence and self-gravity of the interstellar gas, constitutes an attractive base for a unified theory of star formation that is able to explain and reproduce many of the observed features in Galactic star forming regions (Mac Low & Klessen, 2004). Our current study contributes with a detailed analysis of the angular momentum evolution during collapse.

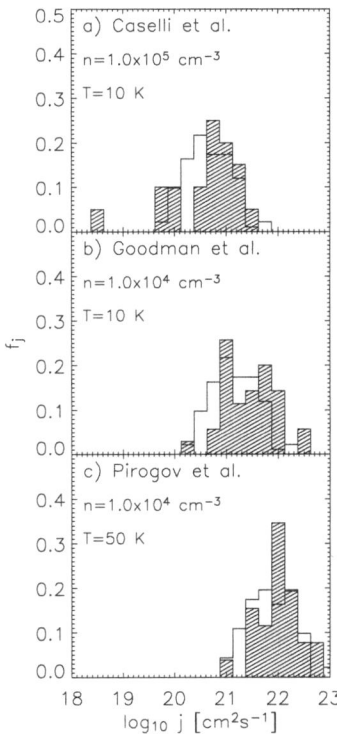

Figure 5.1: The distribution of specific angular momenta of prestellar cores formed in our simulations using model M6k2 (non-hatched histogram) is compared to the distribution of specific angular momenta of observed molecular cloud cores (hatched distributions). The observational data were taken in (a) from Table 5 in Caselli et al. (2002), in (b) from Table 2 in Goodman et al. (1993), Table 4 in Barranco & Goodman (1998) and Table A2 in Jijina et al. (1999) and in (c) from Table 7 in Pirogov et al. (2003). We take f_j to represent the percentage of the total number of existing cores in a specific angular momentum bin. We also give mean number densities n and mean temperatures T for the observations. We use these values to scale the specific angular momenta in the simulations from dimensionless to physical units (for details see text). From Jappsen & Klessen (2004).

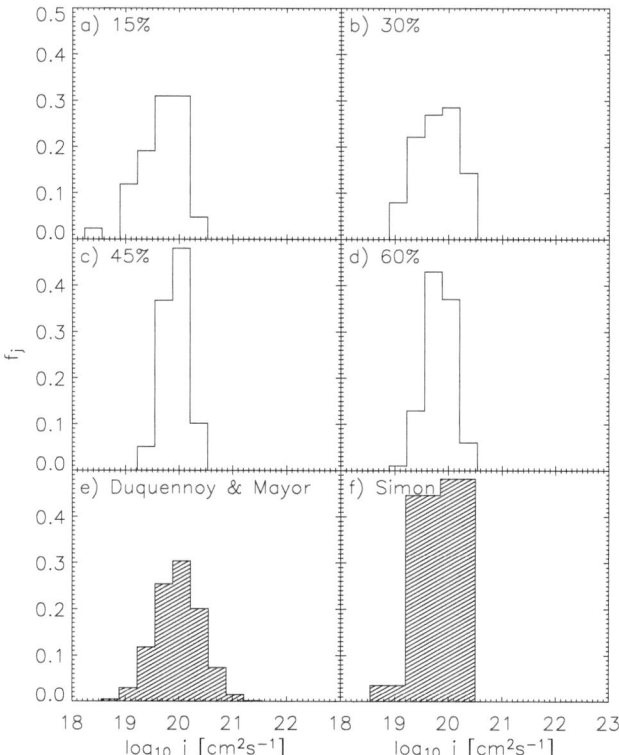

Figure 5.2: Distribution of specific angular momenta of the protostars or protostellar systems at different evolutionary phases of the numerical model M6k2 as denoted by the local star formation efficiency in percent ((a)-(d)). We compare with the j-distribution of binaries among nearby G-dwarf stars from Duquennoy & Mayor (1991) (for details see text) in (e) and with the distribution of specific angular momenta of binaries in the Taurus star-forming region from Figure 5 in Simon (1992) in (f). Again, f_j represents the distribution function which is normalized to 1. From Jappsen & Klessen (2004).

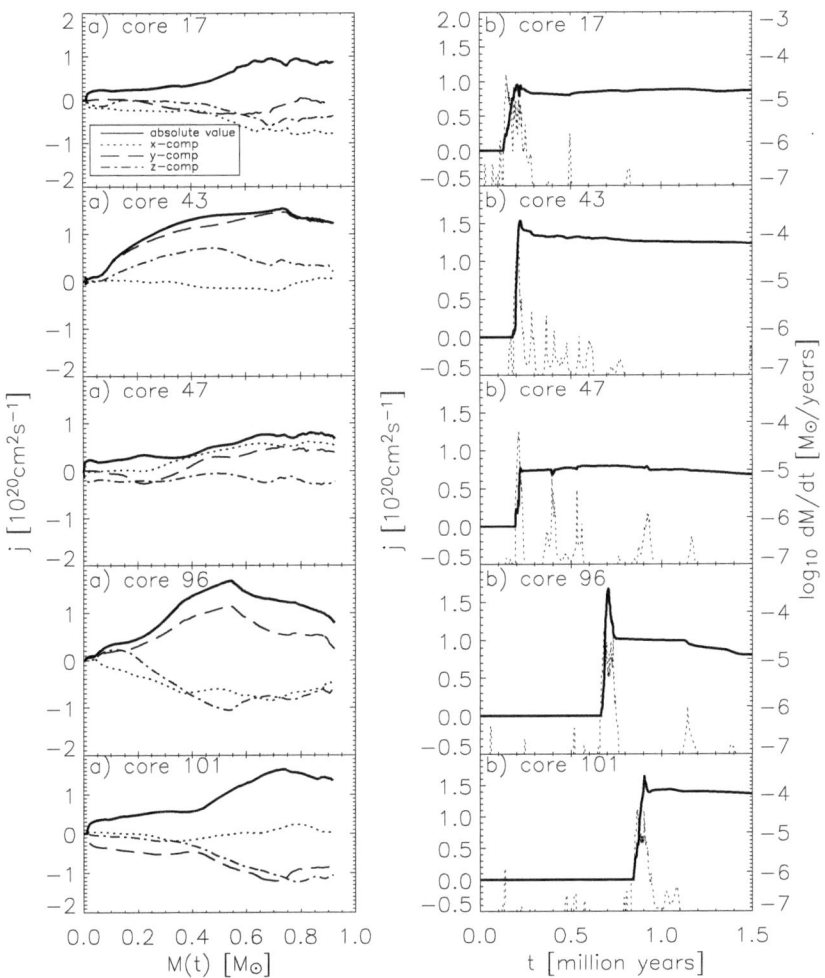

Figure 5.3: Absolute value of specific angular momentum (solid line) of our model M6k2 as (a) a function of mass and as (b) a function of time for five different protostellar objects with approximately equal final masses ($M = 0.94\,M_\odot$). In (a) the x-component (dotted line), the y-component (dashed line) and the z-component (dashed-dotted line) of the specific angular momenta are shown as well. For comparison the mass accretion rates onto the protostar are indicated in (b) by a dashed line (associated y-axis on the right hand side). From Jappsen & Klessen (2004)

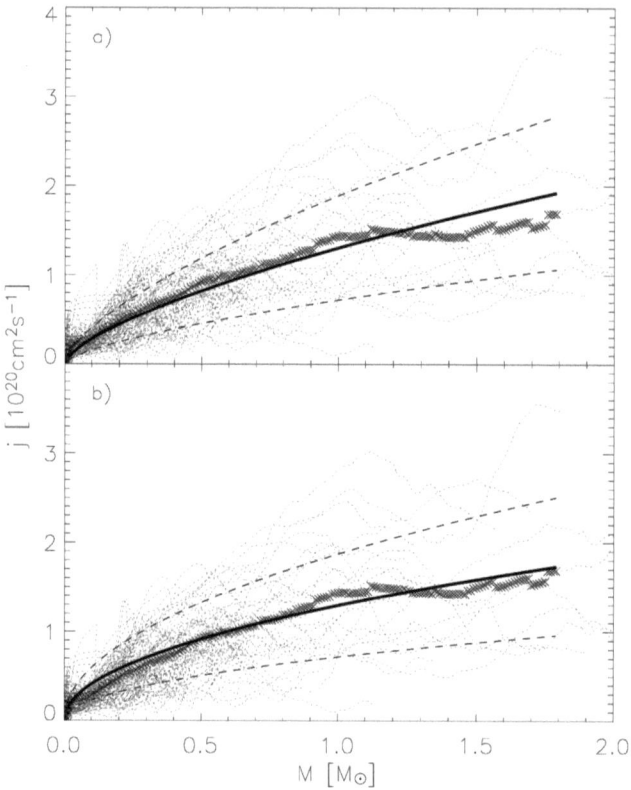

Figure 5.4: Absolute values of specific angular momenta (dotted lines) of all protostars (i.e. sink particles) from our model M6k2 as a function of mass. The specific angular momenta are averaged at certain mass values which are separated by $0.01\,M_\odot$ and the resulting points are indicated by crosses. The solid line represents a fit of these averaged specific angular momenta in the mass range between 0 and $1.7\,M_\odot$. In (a) we fit with a function of the form: $j = \mathcal{A}(M/M_\odot)^{2/3}$, $\mathcal{A} = (1.3 \pm 0.6) \times 10^{20}\,\mathrm{cm^2\,s^{-1}}$, and in (b) we use a square root function: $j = \mathcal{B}\sqrt{M/M_\odot}$, $\mathcal{B} = (1.3 \pm 0.6) \times 10^{20}\,\mathrm{cm^2\,s^{-1}}$. One standard deviation is marked by the dashed lines. From Jappsen & Klessen (2004).

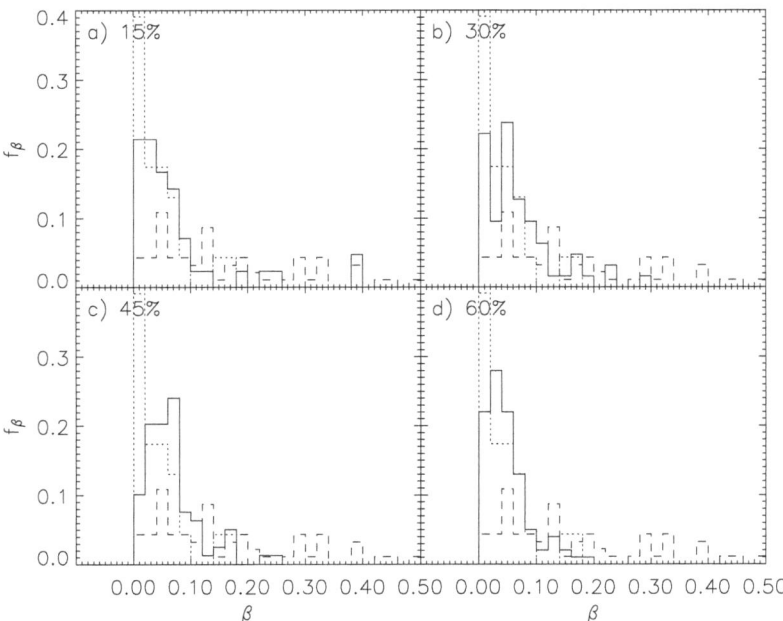

Figure 5.5: Distribution of β obtained from our model M6k2 with 15 % (a), 30 % (b), 45 % (c) and 60 % (d) of the material accreted. For the collapsed protostellar cores in our models (solid lines) we calculate β using Equation 5.6, where we assume solid-body rotation of a uniform density sphere and take the calculated specific angular momentum j. For the prestellar cores (i.e. Jeans-unstable gas clumps; see the dashed lines) we derive β self-consistently from the three-dimensional density and velocity structure (using Equation 5.10). For comparison we also indicate with dotted lines the values reported by Goodman et al. (1993, see their Figure 11) for observed molecular cloud cores. These were obtained with the same assumptions, and we use the same binning with f_β representing the fraction of objects per β bin. From Jappsen & Klessen (2004).

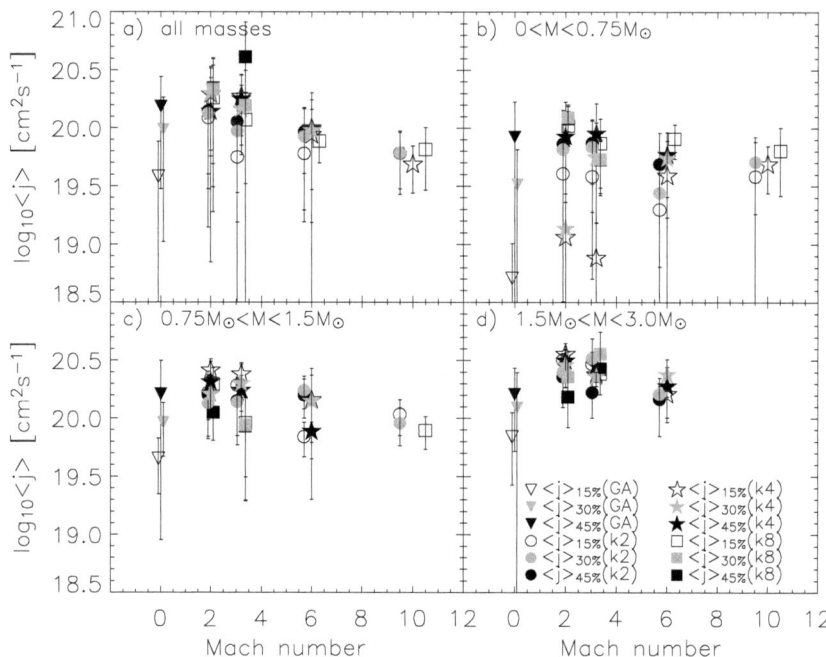

Figure 5.6: Average specific angular momentum of protostellar objects in different turbulent environments as a function of the associated Mach number. Different shapes mark different driving scales k (circle - $k = 2.0$, star - $k = 4.0$, square - $k = 8.0$). GA stands for the Gaussian collapse without driving (triangle pointing downward). Different shades represent different stages of accretion (white - 15% material accreted, gray - 30% material accreted, black - 45% material accreted). In a) all protostars (identified as sink particle in the simulations) were used in calculating the average, in b) through d) only objects in the denoted mass bins were considered. The error bars show the Poissonian standard deviation of $<j>$. For more clarity the symbols are distributed around the corresponding Mach number $\mathcal{M}$ with $\Delta\mathcal{M}/\mathcal{M} = 5\%$. From Jappsen & Klessen (2004).

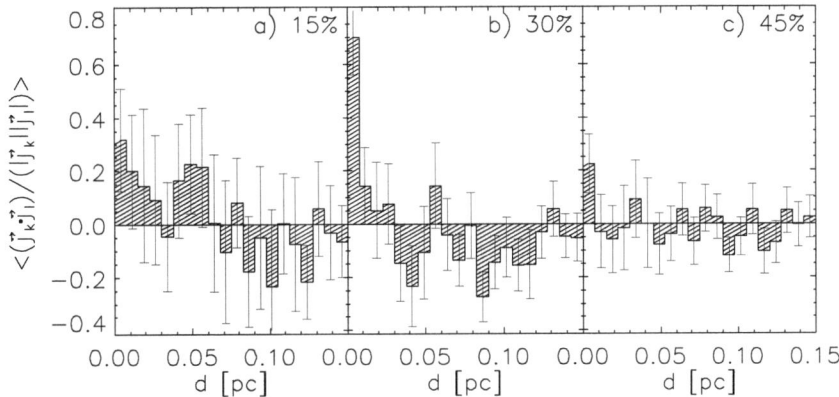

Figure 5.7: The correlation of specific angular momenta of different protostellar cores in model M6k2 with respect to their orientations as a function of distance between the cores. As a measure for the correlation the scalar product of j of different cores was taken and averaged over cores that exhibit similar distances between each other. High positive values denote co-aligned and high negative values denote anti-aligned angular momenta. The three graphs (a)-(c) show three different times at which 15% (a), 30% (b) and 45% (c) of the available material was accreted. The error bars show the Poissonian standard deviation of the averaged correlations. From Jappsen & Klessen (2004).

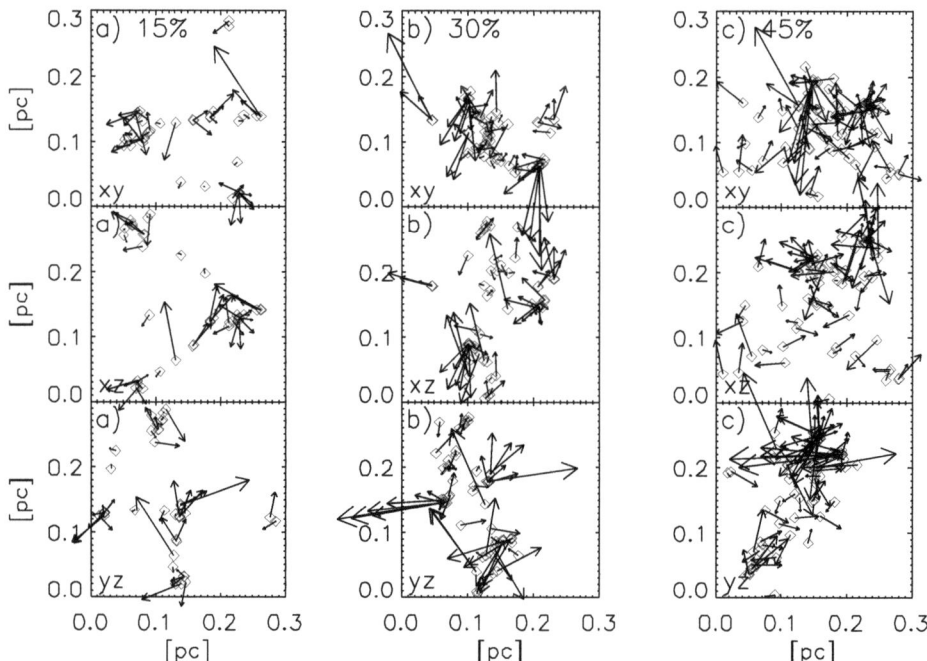

Figure 5.8: Orientation of the angular momenta (arrows) and spatial distribution of the protostellar cores (diamonds) that formed in model M6k2. We present the projections on the xy, xz and yz-plane. The spatial distributions are compared for three different times at which 15% (a), 30% (b) and 45% (c) of the available material is accreted. The length of the arrows scales with the specific angular momentum of the protostellar core. From Jappsen & Klessen (2004).

Chapter 6

Non-isothermal Gravoturbulent Fragmentation

Gravity in galactic molecular clouds is initially expected to be opposed mainly by a combination of supersonic turbulence and magnetic fields (Mac Low & Klessen, 2004). The velocity structure in the clouds is always observed to be dominated by large-scale modes (Mac Low & Ossenkopf, 2000; Ossenkopf et al., 2001; Ossenkopf & Mac Low, 2002). In order to maintain turbulence for some global dynamical timescales and to compensate for gravitational contraction of the cloud as a whole, kinetic energy input from external sources seems to be required. Star formation then takes place in molecular cloud regions which are characterized by local dissipation of turbulence and loss of magnetic flux, eventually leaving thermal pressure as the main force resisting gravity in the small dense prestellar cloud cores that actually build up the stars (Klessen et al., 2005; Vázquez-Semadeni et al., 2005). In agreement with this expectation, observed prestellar cores typically show a rough balance between gravity and thermal pressure (Benson & Myers, 1989; Myers et al., 1991). Therefore the thermal properties of the dense star-forming regions of molecular clouds must play an important role in determining how these clouds collapse and fragment into stars.

In this chapter, we focus on the thermodynamic state of the star-forming gas and how it influences its fragmentation behavior. We also study the role that the thermal properties play in determining the IMF. We address the issue by studying the effects of a piecewise polytropic EOS (see Section 3.4.2) on the formation of stellar clusters in turbulent, self-gravitating molecular clouds using three-dimensional, smoothed particle hydrodynamics simulations. In this completely new set of simulations stars form via a process we call gravoturbulent fragmentation as described in Section 3.3. To approximate the results of published predictions of the thermal behavior of collapsing clouds, we increase the polytropic exponent γ from 0.7 to 1.1 at a critical density n_c, which we estimated to be $2.5 \times 10^5 \, \text{cm}^{-3}$. The change of thermodynamic state at n_c selects a characteristic mass scale for fragmentation M_{ch}, which we relate to the peak of the observed IMF. A simple scaling argument based on the Jeans mass M_J at the critical density n_c leads to $M_{\text{ch}} \propto n_c^{-0.95}$. We perform simulations with $4.3 \times 10^4 \, \text{cm}^{-3} < n_c < 4.3 \times 10^7 \, \text{cm}^{-3}$ to test this scaling argument. Our simulations qualitatively support this hypothesis, but we find a weaker density dependence of $M_{\text{ch}} \propto n_c^{-0.5 \pm 0.1}$. We also investigate the influence of additional environmental parameters on the IMF. We consider variations in the turbulent driving scheme, and consistently find that M_{ch} is decreasing with increasing n_c. Our investigation generally supports the idea that the distribution of stellar masses depends critically on the thermodynamic state of the star-forming gas. The thermodynamic state of interstellar gas is a result of the balance between heating and cooling processes, which in turn are determined by fundamental atomic and molecular physics and by chemical abundances. Given the abundances, the derivation of a characteristic stellar mass can thus be based on universal quantities and constants.

6.1 Model Parameters

In all our models we adopt an initial temperature of 11.4 K corresponding to a sound speed $c_s = 0.2\,\mathrm{km\,s^{-1}}$, a molecular weight μ of 2.36 and an initial number density of $n = 8.4 \times 10^4\,\mathrm{cm^{-3}}$, which is typical for star-forming molecular cloud regions (e.g. ρ-Ophiuchi, see Motte et al., 1998, or the central region of the Orion Nebula Cluster, see Hillenbrand, 1997; Hillenbrand & Hartmann, 1998). Our simulation cube holds a mass of $120\,M_\odot$ and has a size of $L = 0.29\,\mathrm{pc}$. The cube is subject to periodic boundary conditions in every direction (see Section 4.6). The mean initial Jeans mass is $\langle M_\mathrm{J} \rangle_i = 0.7\,M_\odot$.

We use the EOS described in Section 3.4.2, and compute models with $4.3 \times 10^4\,\mathrm{cm^{-3}} \leq n_c \leq 4.3 \times 10^7\,\mathrm{cm^{-3}}$ in four steps, separated by a decade. Note, that the lowest and the highest of these critical densities represent rather extreme cases. From Figure 6.1, where we show the temperature as a function of number density, it is evident that they result in temperatures that are too high or too low compared to observations and theoretical predictions. Nevertheless, including these cases helps us to clarify the influence of a piecewise polytropic EOS. Each simulation starts with a uniform density. Driving begins immediately, while self-gravity is turned on at $t = 2.0\,t_\mathrm{ff}$, after turbulence is fully established. The global free-fall timescale is $t_\mathrm{ff} \approx 10^5\,\mathrm{yr}$. Our models are named mnemonically. R5 up to R8 stand for the critical density n_c ($4.3 \times 10^4\,\mathrm{cm^{-3}} \leq n_c \leq 4.3 \times 10^7\,\mathrm{cm^{-3}}$) in the equation of state, k2 or k8 stand for the wave numbers ($k = 1..2$ or $k = 7..8$) at which the driving energy is injected into the system and b flags the runs with 1 million gas particles. The letter L marks the high resolution runs for critical densities $n_c = 4.3 \times 10^6\,\mathrm{cm^{-3}}$ and $n_c = 4.3 \times 10^7\,\mathrm{cm^{-3}}$ with 2 million and 5.2 million particles, respectively. Different realizations of the turbulent velocity field are denoted by r1, r2, r3. For comparison we also run isothermal simulations marked with the letter I that have particle numbers of approximately 200000, 1 million and 10 million gas particles.

The number of particles determines the minimum resolvable Jeans mass in our models (see Equation 4.7). Figure 3.1 shows the dependence of the local Jeans mass on the density. At the critical density the dependency of the Jeans mass on density changes its behavior. The minimum Jeans mass M_res that needs to be resolved occurs at the density at which sink particles are formed. A local Jeans mass is considered resolved if it contains at least $2 \times N_\mathrm{neigh} = 80$ SPH particles (Bate & Burkert, 1997). As can be seen in Figure 3.1 we are able to resolve M_res with 1 million particles for critical densities up to $n_c = 4.3 \times 10^5\,\mathrm{cm^{-3}}$. Since this is not the case for $n_c = 4.3 \times 10^6\,\mathrm{cm^{-3}}$ and $n_c = 4.3 \times 10^7\,\mathrm{cm^{-3}}$, we repeat these simulations with 2 million and 5.2 million particles, respectively. Due to long calculation times we follow the latter only to the point in time when about 30% of the gas has been accreted.

At the density where γ changes from below unity to above unity, the temperature reaches a minimum. This is reflected in the "V" shape shown in Figure 6.1. All our simulations start with the same initial conditions in temperature and density as marked by the dotted lines. In a further set of simulations we analyze the influence of changing the turbulent driving scheme on fragmentation while using a polytropic equation of state. These models contain 2×10^5 particles each.

Following Bate & Burkert (1997), the runs for $n_c \geq 4.3 \times 10^5$ are not considered fully resolved at the density of sink particle creation, since M_J falls below the critical mass of 80 SPH particles. We note, however, that the global accretion history is not strongly affected and that we only compare similarly unresolved runs. First, we study the effect of different realizations of the turbulent driving fields on typical masses of protostellar objects. We simply select different random numbers to generate the field while keeping the overall statistical properties the same. This allows us to assess the statistical reliability of our results. These models are labeled from R5..8k2r1 to R5..8k2r3. Second, driving in two different wave number ranges is considered. Most models are driven on large scales ($1 \leq k \leq 2$) but we have run a set of models driven on small scales ($7 \leq k \leq 8$) for comparison.

The main model parameters are summarized in Table 6.1.

6.2 Gravoturbulent Fragmentation in Polytropic Gas

Turbulence establishes a complex network of interacting shocks, where converging flows and shear generate filaments of high density. The interplay between gravity and thermal pressure determines the further dynamics of the gas. Adopting a polytropic EOS (Equation 3.21), the choice of the polytropic exponent plays an important role determining the fragmentation behavior. From Equation 3.29 it is evident that $\gamma = 4/3$ constitutes a critical value. A Jeans mass analysis shows that for three-dimensional structures M_J increases with increasing density if γ is above 4/3. Thus, $\gamma > 4/3$ results in the termination of any gravitational collapse. Also, collapse and fragmentation in filaments depend on the equation of state. The equilibrium and stability of filamentary structures has been studied extensively, beginning with Chandrasekhar & Fermi (1953b), and this work has been reviewed by Larson (1985, 2003). For many types of collapse problems, insight into the qualitative behavior of a collapsing configuration can be gained from similarity solutions (Larson, 2003). For the collapse of cylinders with an assumed polytropic equation of state solutions have been derived by Kawachi & Hanawa (1998), and these authors found that the existence of such solutions depends on the assumed value of γ: similarity solutions exist for $\gamma < 1$ but not for $\gamma > 1$. These authors also found that for $\gamma < 1$, the collapse becomes slower and slower as γ approaches unity from below, asymptotically coming to a halt when $\gamma = 1$. This result shows in a particularly clear way that $\gamma = 1$ is a critical case for the collapse of filaments. Kawachi & Hanawa (1998) suggested that the slow collapse that is predicted to occur for γ approaching unity will in reality cause a filament to fragment into clumps, because the timescale for fragmentation then becomes shorter than the timescale for collapse toward the axis of an ideal filament. If the effective value of γ increases with increasing density as the collapse proceeds, as is expected from the predicted thermal behavior discussed in

Table 6.1: Sample parameters, name of the used in the text, driving scale k, critical density n_c, number of SPH particles, number $\mathcal{N}$ of protostellar objects (i.e., "sink particles" in the centers of protostellar cores) at final stage of the simulation, percentage of accreted mass at final stage $M_{\text{acc}}/M_{\text{tot}}$

Name	k	$\log_{10} n_c$ [cm^{-3}]	particle number	$\mathcal{N}$	$\frac{M_{\text{acc}}}{M_{\text{tot}}}$ [%]
Ik2	1..2	—	205 379	59	56
Ik2b	1..2	—	1 000 000	73	78
Ik2L	1..2	—	9 938 375	6	4
R5k2	1..2	4.63	205 379	22	73
R5k2b	1..2	4.63	1 000 000	22	70
R6k2	1..2	5.63	205 379	64	93
R6k2b	1..2	5.63	1 000 000	54	61
R7k2	1..2	6.63	205 379	122	84
R7k2b	1..2	6.63	1 000 000	131	72
R7k2L	1..2	6.63	1 953 125	143	46
R8k2	1..2	7.63	205 379	194	78
R8k2b	1..2	7.63	1 000 000	234	53
R8k2L	1..2	7.63	5 177 717	309	29
R5k8	7..8	4.63	205 379	1	64
R6k8	7..8	5.63	205 379	38	68
R7k8	7..8	6.63	205 379	99	60
R8k8	7..8	7.63	205 379	118	72
R5k2r1	1..2	4.63	205 379	16	62
R6k2r1	1..2	5.63	205 379	34	72
R7k2r1	1..2	6.63	205 379	111	68
R8k2r1	1..2	7.63	205 379	149	64
R5k2r2	1..2	4.63	205 379	21	72
R6k2r2	1..2	5.63	205 379	51	74
R7k2r2	1..2	6.63	205 379	119	70
R8k2r2	1..2	7.63	205 379	184	70
R5k2r3	1..2	4.63	205 379	18	90
R6k2r3	1..2	5.63	205 379	52	85
R7k2r3	1..2	6.63	205 379	123	76
R8k2r3	1..2	7.63	205 379	196	71

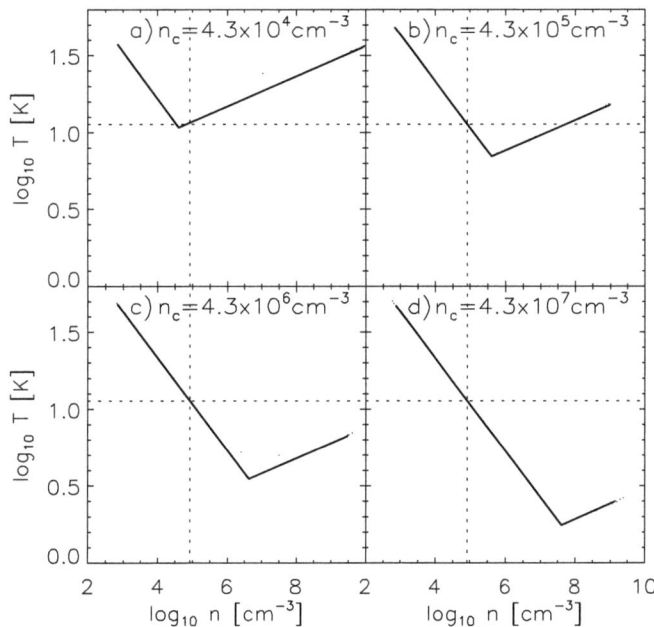

Figure 6.1: Temperature as a function of density for four runs with different critical densities n_c. The dotted lines show the initial conditions. The curve has a discontinuous derivative at the critical density n_c. From Jappsen et al. (2005).

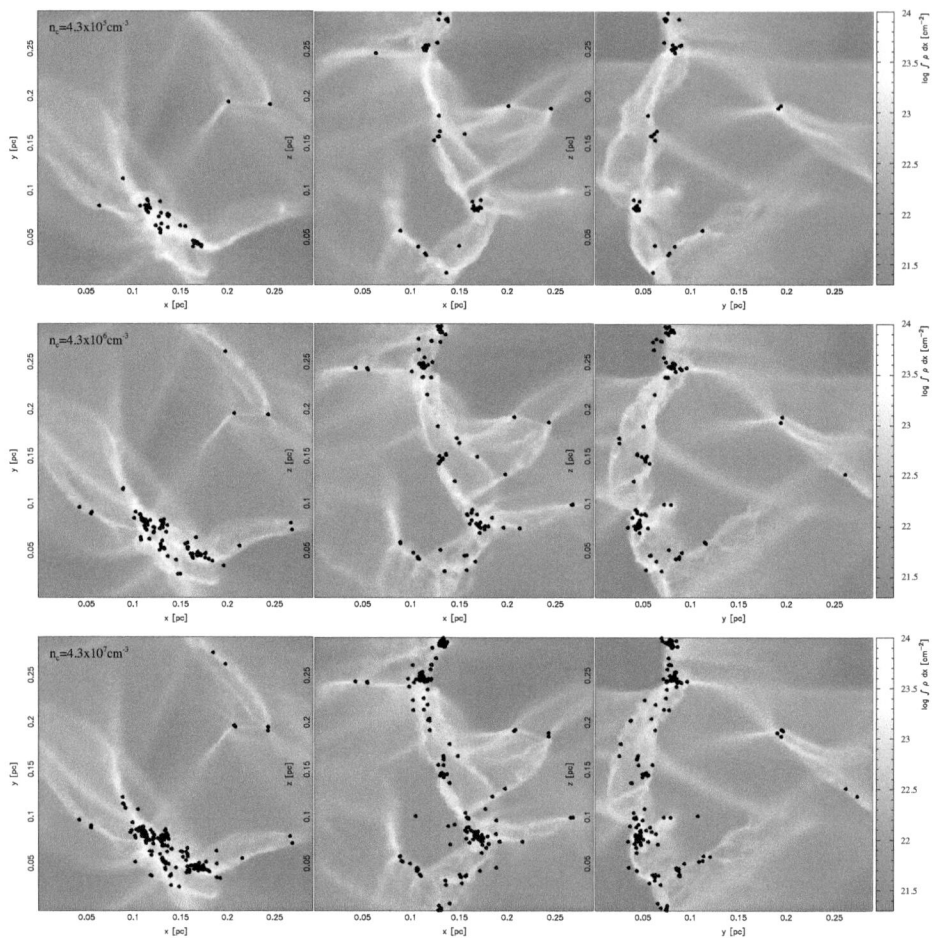

Figure 6.2: Column density distribution of the gas and location of identified protostellar objects (black circles) using the high-resolution models R6..8k2b at the stage where approximately 50% of the gas is accreted. Projections in the xy-, xz-, and yz-plane are shown for three different critical densities. From Jappsen et al. (2005).

Section 2.2, fragmentation may then be particularly favored to occur at the density where γ approaches unity. In their numerical study Li et al. (2003) found, for a range of assumed polytropic equations of state each with constant γ, that the amount of fragmentation that occurs is indeed very sensitive to the value of the polytropic exponent γ, especially for values of γ near unity (see also, Arcoragi et al., 1991).

The fact that filamentary structure is so prominent in our results and other simulations of star formation, together with the fact that most of the stars in these simulations form in filaments, suggests that the formation and fragmentation of filaments must be an important mode of star formation quite generally. This is also supported by the fact that many observed star-forming clouds have filamentary structure, and by the evidence that much of the star formation in these clouds occurs in filaments (Schneider & Elmegreen, 1979; Larson, 1985; Curry, 2002; Hartmann, 2002). As we note in sections 2.2 and 3.4.2 and following Larson (1973b, 2005), the Jeans mass at the density where the temperature reaches a minimum (see Figure 6.1), and hence, γ approaches unity, is predicted to be about $0.3\,M_\odot$, coincidentally close to the mass at which the stellar IMF peaks (also see Figure 3.1). This similarity is an additional hint that filament fragmentation with a varying polytropic exponent may play an important role in the origin of the stellar IMF and the characteristic stellar mass.

The filamentary structure that occurs in our simulations is visualized in Figure 6.2. Here we show the column density distribution of the gas and the distribution of protostellar objects. We display the results for three different critical densities in xy-, xz- and yz-projection. The volume density is computed from the SPH kernel in 3D and then projected along the three principal axes. Figure 6.2 shows for all three cases a remarkably filamentary structure. These filaments define the loci where most protostellar objects form.

Clearly, the change of the polytropic exponent γ at a certain critical density influences the number of protostellar objects formed. If the critical density increases then more protostellar objects form but their mean mass decreases. We show this quantitatively in Figure 6.3. In (a) we compare the number of protostellar objects for different critical densities n_c for the models R5...8k2b. The rate at which new protostars form changes with different n_c. Models that switch from low γ to high γ at low densities built up protostellar objects more rarely than models that change γ at higher densities.

Figure 6.3b shows the accretion histories (the time evolution of the combined mass fraction of all protostellar objects) for the models R5..8k2b. Accretion starts for all but one case approximately at the same time. In model R5k2b, $\gamma = 1.1$ already at the mean initial density, thus γ does not change during collapse. In this case accretion starts at a later time. This confirms the finding by Li et al. (2003) that accretion is delayed for large γ. In the other four cases the accretion history is very similar and the slope is approximately the same for all models.

In both plots we also show the results from our high resolution runs R7k2L and R8k2L. These simulations with 2 million and 5.2 million particles, respectively, have an accretion history similar to the time evolution of the accreted mass fraction in the runs with 1 million particles. The number of protostellar objects, however, is larger for the runs with increased particle numbers. Combining our results in these two figures we find that an environment where γ changes at higher densities produces more, but less massive objects. Thus, the mean mass of protostellar objects does indeed depend on the critical density at which γ changes from 0.7 to 1.1.

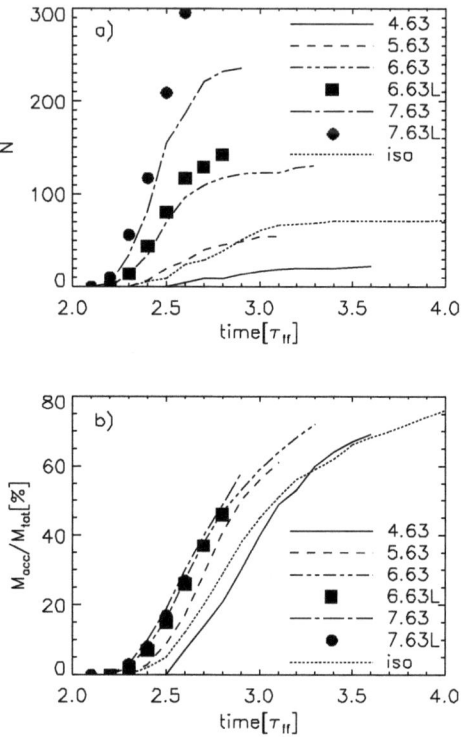

Figure 6.3: Temporal evolution of the number of protostellar objects (upper plot) and of the ratio of accreted gas mass to total gas mass (lower plot) for models R5..8k2b. The legend shows the logarithms of the respective number densities in cm^{-3}. The times are given in units of a free-fall time τ_{ff}. We also show the models R7k2L and R8k2L with 2×10^6 and 5×10^6 particles, respectively, which are denoted by the letter "L". For comparison the dotted lines indicate the values for the isothermal model Ik2b. From Jappsen et al. (2005).

6.3 Dependence of the Characteristic Mass on the Equation of State

Further insight into how the characteristic stellar mass may depend on the critical density can be gained from the mass spectra of the protostellar objects, which we show in Figure 6.4. We plot the mass spectra of models R5...6k2b, model R7k2L and model R8k2L at different times, when the fraction of mass accumulated in protostellar objects has reached approximately $10\%, 30\%$ and 50%. In the top row we also display the results of an isothermal run for comparison. We used the same initial conditions and parameters in all models shown. Dashed lines indicate the mass resolution limits.

We find closest correspondence with the observed IMF (see, Scalo, 1998; Kroupa, 2002; Chabrier, 2003) for a critical density of $4.3 \times 10^6 \, \mathrm{cm}^{-3}$ and for stages of accretion around 30% and above. At high masses, our distribution follows a Salpeter-like power law. For comparison we indicate the Salpeter slope $x \approx 1.3$ (Salpeter, 1955) where the IMF is defined by $dN/d\log m \propto m^{-x}$ (see Equation 2.2). For masses close to the median mass the distribution exhibits a small plateau and then falls off towards smaller masses.

The model R5k2b where the change in γ occurs below the initial mean density, shows a flat distribution with only few, but massive protostellar objects. They reach masses up to $10 \, M_\odot$ and the minimum mass is about $0.3 \, M_\odot$. All other models build up a power-law tail towards high masses. This is due to protostellar accretion processes, as fragmentation starts earlier and eventually more and more gas gets turned into stars (see also, Bonnell et al., 2001b; Klessen, 2001; Schmeja & Klessen, 2004). The distribution becomes more peaked for higher n_c and there is a shift of the peak to lower masses. This is already visible in the mass spectra when the protostellar objects have only accreted 10% of the total mass. Model R8k2L has minimum and maximum masses of $0.013 \, M_\odot$ and $1.0 \, M_\odot$, respectively. There is a gradual shift in the median mass (as indicated by the vertical line) from Model R5k2b, with $M_{\mathrm{med}} = 2.5 \, M_\odot$ at 30%, to Model R8k2L, with $M_{\mathrm{med}} = 0.05 \, M_\odot$ at 30%. A similar trend is noticeable during all phases of the model evolution.

What happens when γ increases above unity at the critical density ρ_c? One suggestion is that the increase in γ is sufficient to strongly reduce fragmentation at higher densities, introducing a characteristic scale into the mass spectrum at the value of the Jeans mass at ρ_c. Then the behavior of the Jeans mass with increasing critical density would immediately allow us to derive the scaling law

$$M_{\mathrm{ch}} \propto \rho_c^{-0.95}. \tag{6.1}$$

This simple analytical consideration would then predict a characteristic mass scale which corresponds to a peak of the IMF at $0.35 \, M_\odot$ for a critical density of $\rho_c = 10^{-18} \, \mathrm{g\,cm}^{-3}$ or equivalently a number density of $n_c = 2.5 \times 10^5 \, \mathrm{cm}^{-3}$ when using a mean molecular weight $\mu = 2.36$ appropriate for solar metallicity molecular clouds in the Milky Way. Note, however, that this simple scaling law does not take any further dynamical processes into account.

This change of median mass with critical density n_c is depicted in Figure 6.5. Again, we consider models R5...6k2b, model R7k2L and model R8k2L. The median mass decreases clearly with increasing critical density as expected. As we resolve higher density contrasts the median collapsed mass decreases. We fit our data with a straight lines. The slopes take values between -0.4 and -0.6. These values are larger than the slope -0.95 derived from

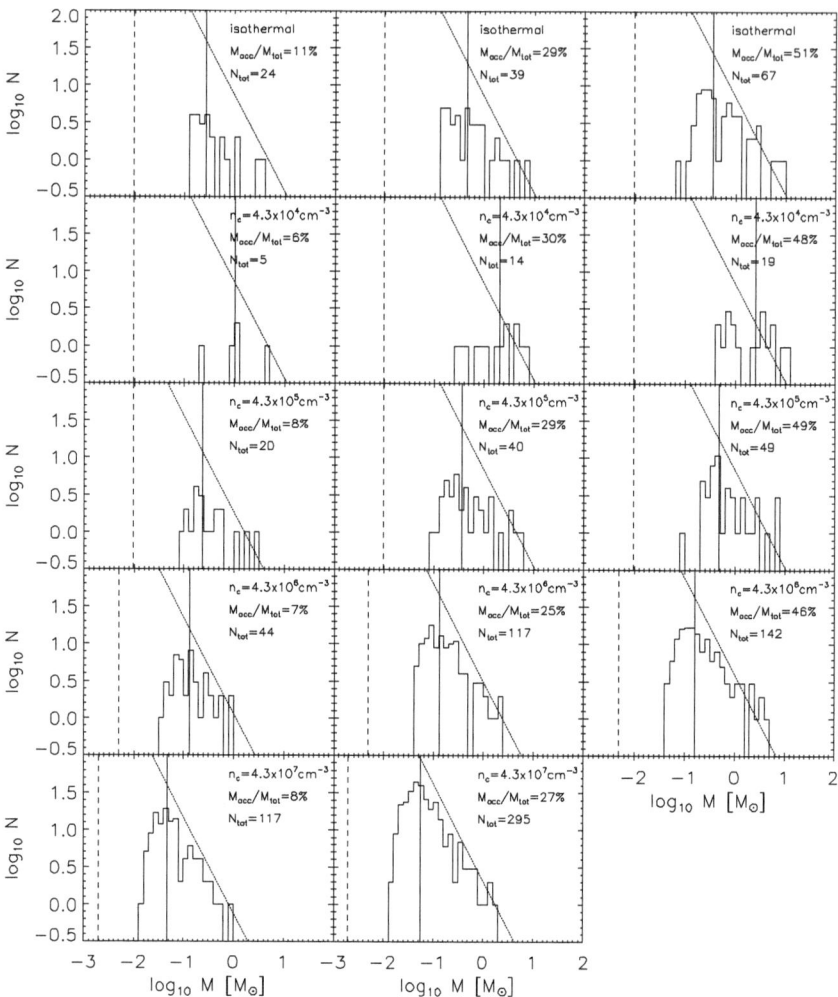

Figure 6.4: Mass spectra of protostellar objects for models R5..6k2b, model R7k2L and model R8k2L at 10%, 30% and 50% of total mass accreted on these protostellar objects. For comparison we also show in the first row the mass spectra of the isothermal run Ik2b. Critical density n_c, ratio of accreted gas mass to total gas mass M_{acc}/M_{tot} and number of protostellar objects are given in the plots. The vertical solid line shows the position of the median mass. The dotted line has a slope of -1.3 and serves as a reference to the Salpeter value (Salpeter, 1955). The dashed line indicates the mass resolution limit. Model R8k2L was stopped at 30% due to long computational time. From Jappsen et al. (2005).

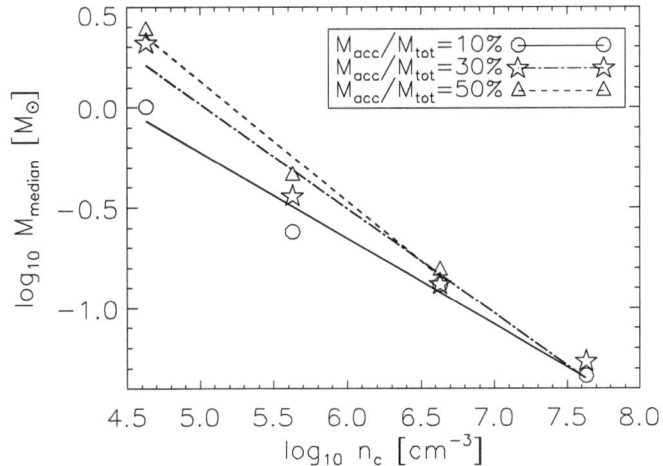

Figure 6.5: The plot shows the median mass of the protostellar objects over critical density for models R5..6k2b, model R7k2L and model R8k2L. We display results for different ratios of accreted gas mass to total gas mass $M_{\rm acc}/M_{\rm tot}$. We fit our data with straight lines for different stages of accretion. The slopes have the following values: -0.43 ± 0.05 (solid line), -0.52 ± 0.06 (dashed-dotted line), -0.60 ± 0.07 (dashed line). From Jappsen et al. (2005).

the simple scaling law (Equation 6.1) based on calculation of the Jeans mass $M_{\rm J}$ at the critical density $n_{\rm c}$.

One possible reason for this deviation is the fact that most of the protostellar objects are members of tight groups. Hence, they are subject to mutual interactions and competitive accretion that may change the environmental context for individual protostars. This in turn influences the mass distribution and its characteristics (see, e.g. Bonnell et al., 2001a,b). Another possible reason is that the mass that goes into filaments and then into collapse may depend on further environmental parameters, some of which we discuss in Section 6.4.

6.4 Dependence of the Characteristic Mass on Environmental Parameters

6.4.1 Dependence on Realization of the Turbulent Velocity Field

We compare models with different realizations of the turbulent driving field in Figure 6.6. We fit our data with straight lines for each stage of accretion. Figure 6.6a shows the results of models R5..8k2 which were calculated with the same parameters but lower resolution than the models used for Figure 6.5. As discussed in Section 6.2, although the number of protostellar objects changes with the number of particles in the simulation, the time

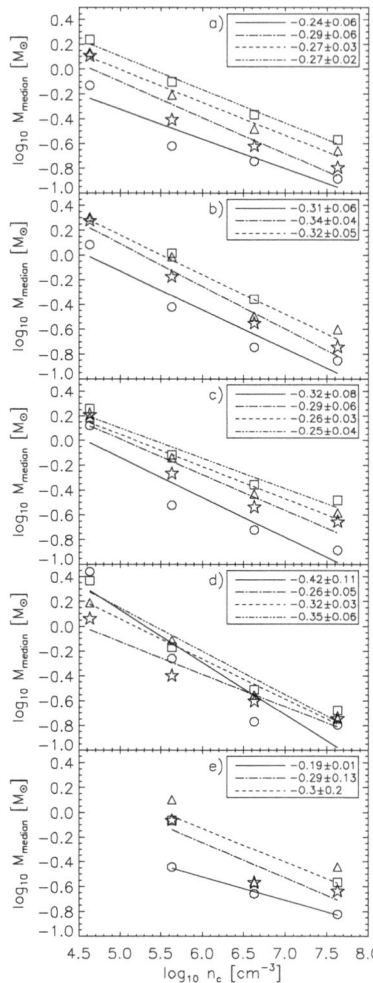

Figure 6.6: Median mass of protostellar objects over critical density at different evolutionary phases (ratio of accreted gas mass over total gas mass M_{acc}/M_{tot}:10% (circle), 30% (star), 50% (triangle), 70% (square)). In (a)-(d) we show identical models but with different realizations of the turbulent velocity field ((a): R5..8k2, (b): R5..8k2r1, (c): R5..8k2r2, (d): R5..8k2r3). The models R6..8k8 in (e) are driven on a smaller scale $k = 7..8$ than in the other cases. All relevant parameters are summarized in Table 6.1. We fit the median values with straight lines for different stages of accretion. The slopes are given in the plot and denoted as follows: 10% – solid line, 30% – dashed-dotted line, 50% – dashed line, 70% – dashed-double-dotted line. From Jappsen et al. (2005).

evolution of the total mass accreted on all protostellar objects remains similar. Thus, lower resolution models exhibit the same general trend as their high-resolution counterparts and show the same global dependencies. We notice, however, that the slope of the M_{median}-n_{c} relation typically is shallower in the low-resolution models. This can be seen when comparing Figure 6.5 and Figure 6.6a, where we used identical turbulent driving fields. This is clearly a numerical artifact because more objects form artificially in low resolution runs.

For all four different realizations of the turbulent driving field shown in Figure 6.6a-d we see a clear trend of decreasing median mass M_{median} with increasing n_{c}. We conclude that, qualitatively, the M_{median}-n_{c} relation is independent of the details of the turbulent driving field but, quantitatively, there are significant variations. This is not surprising given the stochastic nature of turbulent flows. A further discussion on this issue can be found in Klessen et al. (2000) and Heitsch et al. (2001).

6.4.2 Dependence on the Scale of Turbulent Driving

Figure 6.6e shows the results for models where the driving scale has been changed to a lower value ($k = 7..8$). The overall dependency of M_{median} on n_{c} is very similar to the cases of large-scale turbulence. However, we note considerably larger uncertainties in the exact value of the slope. This holds especially for the phases where 30% and 50% of the total gas mass has been converted into stars. One of the reasons is lower statistics, i.e. for the smallest critical density only one protostellar object forms. Moreover, it has already been noted by Klessen (2001); Li et al. (2003) that driving on small wavelength results in less fragmentation. The small-scale density structure formed is not so strongly filamentary, compared to the case of large-scale driven turbulence. Local differences have a larger influence on the results for driving on small wavelengths. Nevertheless, for most of the models the mean mass decreases with increasing critical density. Observational evidence suggests that real molecular clouds are driven mainly from large scales (e.g. Ossenkopf & Mac Low, 2002; Brunt et al., 2005), although some authors claim that jets and winds provide driving from small scales (Shang et al., 2006).

6.5 Summary

Using SPH simulations we investigate the influence of a piecewise polytropic EOS on fragmentation of molecular clouds. We study the case where the polytropic index γ changes from a value below to one above unity at a critical density n_{c}. We consider a broad range of values of n_{c} around a realistic value to determine the dependence of the mass spectrum on n_{c}.

Observational evidence predicts that dense prestellar cloud cores show a rough balance between gravity and thermal pressure (Benson & Myers, 1989; Myers et al., 1991). Thus, the thermodynamical properties of the gas play an important role in determining how dense star-forming regions in molecular clouds collapse and fragment. Observational and theoretical studies of the thermal properties of collapsing clouds both indicate that at densities below 10^{-18} g cm^{-3}, roughly corresponding to a number density of $n_c = 2.5 \times 10^5$ cm^{-3}, the temperature decreases with increasing density. This is due to the strong dependence of molecular cooling rates on density (Koyama & Inutsuka, 2000). Therefore, the polytropic exponent γ is below unity in this density regime. At densities above 10^{-18} g cm^{-3}, the gas becomes thermally coupled to the dust grains, which then control the gas temperature by far-infrared thermal continuum emission. The balance between compressional heating and thermal cooling by dust causes the temperature to increase again slowly with increasing density. Thus the temperature-density relation can be approximated with γ above unity (Larson, 1985) in this regime. Changing γ from a value below unity to a value above unity results in a minimum temperature at the critical density. Li et al. (2003) showed that gas fragments efficiently for constant $\gamma < 1.0$ and less efficiently for higher γ or not not all if gas distribution is filamentary and $\gamma > 1$. Thus, the Jeans mass at the critical density defines a characteristic mass for fragmentation, which may be related to the peak of the IMF.

We investigate this relation numerically by changing γ from 0.7 to 1.1 at different critical densities n_c varying from 4.3×10^4 cm^{-3} to 4.3×10^7 cm^{-3}. A simple scaling argument based on the Jeans mass M_J at the critical density n_c leads to $M_J \propto n_c^{-0.95}$. If there is a close relation between the average Jeans mass and the characteristic mass of a fragment, a similar relation should hold for the expected peak of the mass spectrum. Our simulations qualitatively support this hypothesis, however, with the weaker density dependency $M_{ch} \propto n_c^{-0.5 \pm 0.1}$. The density at which γ changes from below unity to above unity selects a characteristic mass scale. Consequently, the peak of the resulting mass spectrum decreases with increasing critical density. This spectrum not only shows a pronounced peak but also a power-law tail towards higher masses. Its behavior is thus similar to the observed IMF.

Altogether, supersonic turbulence in self-gravitating molecular gas generates a complex network of interacting filaments. The overall density distribution is highly inhomogeneous. Turbulent compression sweeps up gas in some parts of the cloud, but other regions become rarefied. The fragmentation behavior of the cloud and its ability to form stars depend strongly on the EOS. If collapse sets in, the final mass of a fragment depends not only on the local Jeans criterion, but also on additional processes. For example, protostars grow in mass by accretion from their surrounding material. In turbulent clouds the properties of the gas reservoir are continuously changing. In a dense cluster environment, furthermore, protostars may interact with each other, leading to ejection from the cluster or mass exchange. These dynamical factors modify the resulting mass spectrum, and may explain why the characteristic stellar mass depends on the EOS more weakly than expected.

We also studied the effects of different turbulent driving fields and of a smaller driving scale. For different realizations of statistically identical large-scale turbulent velocity fields we consistently find that the characteristic mass decreases with increasing critical mass. However, there are considerable variations. The influence of the natural stochastic fluctuations in the turbulent flow on the resulting median mass is almost as pronounced as the changes of the thermal properties of the gas. Also when inserting turbulent energy at small wavelengths we see the peak of the mass spectrum decrease with increasing critical density.

Our investigation supports the idea that the distribution of stellar masses depends, at least in part, on the thermodynamic state of the star-forming gas. If there is a low-density regime in molecular clouds where temperature T decreases with increasing density ρ, followed by a higher-density phase where T increases with ρ, fragmentation seems likely to be favored at the transition density where the temperature reaches a minimum. This defines a characteristic mass scale. The thermodynamic state of interstellar gas is a result of the balance between heating and cooling processes, which in turn are determined by fundamental atomic and molecular physics and by chemical abundances. The derivation of a characteristic stellar mass can thus be based on quantities and constants that depend solely on the chemical abundances in a molecular cloud. The current study using a piecewise polytropic EOS can only serve as a first step. Future work will need to consider a realistic chemical network and radiation transfer processes in gas of varying abundances.

Chapter 7

Cooling and Collapse of Ionized Gas in Small Protogalactic Halos

In the preceding chapter we discussed how the thermodynamical behavior of star-forming gas can be modeled with a polytropic EOS. Nevertheless, we also stated that a realistic chemical network and the consideration of radiation transfer processes would be desirable for the study of star formation in gas of varying abundances. Whereas this is still a future goal for molecular clouds in the present universe, we can approach the problem more easily in regions with metal free gas or gas with very low metallicity. The important chemical reactions and cooling processes are low in number and therefore computationally feasible. In particular, we are interested in the formation of the first and mainly the second generation of stars ever formed.

We study the cooling properties and the collapse of ionized gas in small protogalactic dark matter halos (see Section 3.1.2) using three-dimensional, smoothed particle hydrodynamics simulations. Again, we use the publicly available parallel code gadget (Springel et al., 2001) in which we have implemented a sink particle algorithm which allows us to safely represent gas which has collapsed beyond the resolution limit without causing numerical errors within the resolved regions of the simulation. Instead of dark matter particles, we use a fixed background potential to model the influence of a dark matter halo (see Section 7.1). We have also added the necessary framework for following the non-equilibrium chemistry of H_2 in the protogalactic gas and an appropriate treatment of radiative heating and cooling. In contrast to most previous simulations, we have also incorporated self-shielding of H_2, albeit in a rather approximate manner.

Our initial conditions represent protogalaxies forming within what Oh & Haiman (2003) call a "fossil" H ii region - a previously ionized H ii region which has not yet had time to cool and recombine. Prior to cosmological reionization, such regions should be relatively common, since the characteristic lifetime of the likely ionizing sources are significantly shorter than a Hubble time. In this study we investigate how the collapse of the initially ionized gas depends on various parameters such as redshift, metallicity, mass of the dark matter halo and strength of the ultraviolet background.

Table 7.1: Sample parameters of the simulations, name of the simulation, redshift z, ν, mass of the dark matter halo $M_{\rm DM}$, metallicity of the gas Z, UV background J_{21} (see Equation 4.12), mass per SPH particle $m_{\rm part}$, length of the computational box L and the scale radius $r_{\rm s}$.

name	z	ν	$M_{\rm DM}$ [$M_\odot$]	Z [$Z_\odot$]	J_{21}	$m_{\rm part}$ [$M_\odot$]	L [kpc]	$r_{\rm s}$ [kpc]
ZM15s	15	1.7	6.7×10^4	0.0	0.0	0.25	0.7	0.019
ZM20s	20	2.2	5.3×10^4	0.0	0.0	0.25	0.5	0.014
ZM25s	25	2.7	4.5×10^4	0.0	0.0	0.25	0.5	0.011
ZM30s	30	3.2	4.0×10^4	0.0	0.0	0.25	0.5	0.0092
ZM15m	15	2.2	4.2×10^6	0.0	0.0	5.0	2.8	0.076
ZM20m	20	2.7	1.5×10^6	0.0	0.0	5.0	1.6	0.043
ZM25m	25	3.2	7.8×10^5	0.0	0.0	2.5	1.0	0.029
ZM30m	30	3.7	4.8×10^5	0.0	0.0	2.0	0.8	0.021
ZM15l	15	2.4	1.5×10^7	0.0	0.0	5.0	4.2	0.118
ZM20l	20	3.1	1.2×10^7	0.0	0.0	5.0	3.0	0.088
ZM25l	25	3.7	7.0×10^6	0.0	0.0	5.0	2.2	0.061
ZM30l	30	4.3	4.8×10^6	0.0	0.0	5.0	1.8	0.046
ZM15sUV	15	1.7	6.7×10^4	0.0	10^{-2}	0.25	0.7	0.019
ZM20sUV	20	2.2	5.3×10^4	0.0	10^{-2}	0.25	0.5	0.014
ZM25sUV	25	2.7	4.5×10^4	0.0	10^{-2}	0.25	0.5	0.011
ZM30sUV	30	3.2	4.0×10^4	0.0	10^{-2}	0.25	0.5	0.0092
ZM15mUV	15	2.2	4.2×10^6	0.0	10^{-2}	5.0	2.8	0.076
ZM20mUV	20	2.7	1.5×10^6	0.0	10^{-2}	5.0	1.6	0.043
ZM25mUV	25	3.2	7.8×10^5	0.0	10^{-2}	2.5	1.0	0.029
ZM30mUV	30	3.7	4.8×10^5	0.0	10^{-2}	2.0	0.8	0.021
ZM15lUV	15	2.4	1.5×10^7	0.0	10^{-2}	5.0	4.2	0.118
ZM20lUV	20	3.1	1.2×10^7	0.0	10^{-2}	5.0	3.0	0.088
ZM25lUV	25	3.7	7.0×10^6	0.0	10^{-2}	5.0	2.2	0.061
ZM30lUV	30	4.3	4.8×10^6	0.0	10^{-2}	5.0	1.8	0.046

7.1 Initial Conditions

As we wish to run a large number of simulations of protogalactic collapse, we have chosen to limit the computational cost of each simulation by starting from somewhat simplified initial conditions. Since we are not particularly interested (at this stage at least) in following the assembly history of the dark matter halo in which the protogalaxy resides, or in studying the response of the halo to the cooling of the gas, we choose to model the influence of the halo by using a fixed background potential and not by dark matter particles. To construct this potential, we assume that the halo is spherically symmetric, with the density profile of Navarro et al. (1996):

Table 7.1: –Continued

name	z	ν	$M_{\rm DM}$ [M$_\odot$]	Z [$Z_\odot$]	J_{21}	$m_{\rm part}$ [M$_\odot$]	L [kpc]	$r_{\rm s}$ [kpc]
ZM15sUV2	15	1.7	6.7×10^4	0.0	10^{-1}	0.25	0.7	0.019
ZM20sUV2	20	2.2	5.3×10^4	0.0	10^{-1}	0.25	0.5	0.014
ZM25sUV2	25	2.7	4.5×10^4	0.0	10^{-1}	0.25	0.5	0.011
ZM30sUV2	30	3.2	4.0×10^4	0.0	10^{-1}	0.25	0.5	0.0092
ZM15mUV2	15	2.2	4.2×10^6	0.0	10^{-1}	5.0	2.8	0.076
ZM20mUV2	20	2.7	1.5×10^6	0.0	10^{-1}	5.0	1.6	0.043
ZM25mUV2	25	3.2	7.8×10^5	0.0	10^{-1}	2.5	1.0	0.029
ZM30mUV2	30	3.7	4.8×10^5	0.0	10^{-1}	2.0	0.8	0.021
ZM15lUV2	15	2.4	1.5×10^7	0.0	10^{-1}	5.0	4.2	0.118
ZM20lUV2	20	3.1	1.2×10^7	0.0	10^{-1}	5.0	3.0	0.088
ZM25lUV2	25	3.7	7.0×10^6	0.0	10^{-1}	5.0	2.2	0.061
ZM30lUV2	30	4.3	4.8×10^6	0.0	10^{-1}	5.0	1.8	0.046
LM15s	15	1.7	6.7×10^4	10^{-3}	0.0	0.25	0.7	0.019
LM20s	20	2.2	5.3×10^4	10^{-3}	0.0	0.25	0.5	0.014
LM25s	25	2.7	4.5×10^4	10^{-3}	0.0	0.25	0.5	0.011
LM30s	30	3.2	4.0×10^4	10^{-3}	0.0	0.25	0.5	0.0092
LM15m	15	2.2	4.2×10^6	10^{-3}	0.0	5.0	2.8	0.076
LM20m	20	2.7	1.5×10^6	10^{-3}	0.0	5.0	1.6	0.043
LM25m	25	3.2	7.8×10^5	10^{-3}	0.0	2.5	1.0	0.029
LM30m	30	3.7	4.8×10^5	10^{-3}	0.0	2.0	0.8	0.021
LM15l	15	2.4	1.5×10^7	10^{-3}	0.0	5.0	4.2	0.118
LM20l	20	3.1	1.2×10^7	10^{-3}	0.0	5.0	3.0	0.088
LM25l	25	3.7	7.0×10^6	10^{-3}	0.0	5.0	2.2	0.061
LM30l	30	4.3	4.8×10^6	10^{-3}	0.0	5.0	1.8	0.046

$$\rho_{\rm dm}(r) = \frac{\delta_{\rm c} \rho_{\rm crit}}{r/r_{\rm s}(1+r/r_{\rm s})^2}, \qquad (7.1)$$

where $r_{\rm s}$ is a scale radius, $\delta_{\rm c}$ is a characteristic (dimensionless) density and $\rho_{\rm crit} = 3H^2/8\pi G$ is the critical density for closure. Following Navarro et al. (1997) we calculate the characteristic density and scale radius using a given redshift and dark halo mass. We choose halo masses for four different redshifts z: 15, 20, 25 and 30. Numerical studies of the formation of primordial gas clouds and the first stars indicate that baryonic collapse due to H$_2$ began as early as $z \sim 30$ (Abel et al., 2002; Bromm et al., 2002). Thus, we choose this redshift as our highest value and investigate the behavior at redshifts down to 15, as the wmap polarization results (Kogut et al., 2003) suggest that the cosmological reionization occurred somewhere in the redshift range 17±5. According to the Press & Schechter (1974) formalism (see Section 3.1.2) we choose 3 different halo masses from the halo mass function with the parameter ν varying between 1.7 and 4.3. We choose ν to be around 3.0, since such halos are often taken to be representative of the earliest objects to form, although this choice is

Table 7.1: –Continued

name	z	ν	$M_{\rm DM}$ [$M_\odot$]	Z [$Z_\odot$]	J_{21}	$m_{\rm part}$ [$M_\odot$]	L [kpc]	r_s [kpc]
LM15sUV	15	1.7	6.7×10^4	10^{-3}	10^{-2}	0.25	0.7	0.019
LM20sUV	20	2.2	5.3×10^4	10^{-3}	10^{-2}	0.25	0.5	0.014
LM25sUV	25	2.7	4.5×10^4	10^{-3}	10^{-2}	0.25	0.5	0.011
LM30sUV	30	3.2	4.0×10^4	10^{-3}	10^{-2}	0.25	0.5	0.0092
LM15mUV	15	2.2	4.2×10^6	10^{-3}	10^{-2}	5.0	2.8	0.076
LM20mUV	20	2.7	1.5×10^6	10^{-3}	10^{-2}	5.0	1.6	0.043
LM25mUV	25	3.2	7.8×10^5	10^{-3}	10^{-2}	2.5	1.0	0.029
LM30mUV	30	3.7	4.8×10^5	10^{-3}	10^{-2}	2.0	0.8	0.021
LM15lUV	15	2.4	1.5×10^7	10^{-3}	10^{-2}	5.0	4.2	0.118
LM20lUV	20	3.1	1.2×10^7	10^{-3}	10^{-2}	5.0	3.0	0.088
LM25lUV	25	3.7	7.0×10^6	10^{-3}	10^{-2}	5.0	2.2	0.061
LM30lUV	30	4.3	4.8×10^6	10^{-3}	10^{-2}	5.0	1.8	0.046
LM15sUV2	15	1.7	6.7×10^4	10^{-3}	10^{-1}	0.25	0.7	0.019
LM20sUV2	20	2.2	5.3×10^4	10^{-3}	10^{-1}	0.25	0.5	0.014
LM25sUV2	25	2.7	4.5×10^4	10^{-3}	10^{-1}	0.25	0.5	0.011
LM30sUV2	30	3.2	4.0×10^4	10^{-3}	10^{-1}	0.25	0.5	0.0092
LM15mUV2	15	2.2	4.2×10^6	10^{-3}	10^{-1}	5.0	2.8	0.076
LM20mUV2	20	2.7	1.5×10^6	10^{-3}	10^{-1}	5.0	1.6	0.043
LM25mUV2	25	3.2	7.8×10^5	10^{-3}	10^{-1}	2.5	1.0	0.029
LM30mUV2	30	3.7	4.8×10^5	10^{-3}	10^{-1}	2.0	0.8	0.021
LM15lUV2	15	2.4	1.5×10^7	10^{-3}	10^{-1}	5.0	4.2	0.118
LM20lUV2	20	3.1	1.2×10^7	10^{-3}	10^{-1}	5.0	3.0	0.088
LM25lUV2	25	3.7	7.0×10^6	10^{-3}	10^{-1}	5.0	2.2	0.061
LM30lUV2	30	4.3	4.8×10^6	10^{-3}	10^{-1}	5.0	1.8	0.046

somewhat arbitrary. These halos are moderately rare objects, representing no more than a few thousandths of the total cosmic mass (Mo & White, 2002), but are sufficiently common, so that one would expect to find many of them within a single Hubble volume. Table 7.1 gives an overview of the different simulations. The number in the notation of the run name stands for the redshift at the start of the run, whereas s, m, l denote small, medium and large halo masses respectively. We truncate the halo at a radius r_t where $\rho_{\rm dm}$ equals the background density at this redshift. For the gas, we assume an initially uniform distribution, with an initial density ρ_g, taken to be equal to the cosmological background density. We evolve the gas for 1 Gyr until the gas follows the dark matter distribution. The initial temperature of the gas is taken to be uniform, with a value $T_g = 10000\,{\rm K}$, the choice of which is discussed below.

The computational volume is a box of side length L. The length is chosen so that it is the maximum of $2r_t$ and twice the radius at which the gas density is almost uniform. We use periodic boundary conditions for the hydrodynamic part of the force calculations to keep the gas bound within the computational volume. The self-gravity of the gas and the gravitational force exerted by the dark matter potential are not calculated periodically (see Section 4.6).

The quantity of gas present in our simulation was taken to be a fraction Ω_b/Ω_{dm} of the total dark matter (baryon density over dark matter density), where $\Omega_{dm} = \Omega_m - \Omega_b$. We took values for the cosmological parameters from Spergel et al. (2003), and so the baryon density is $\Omega_b = 0.047$ and the matter density is $\Omega_m = 0.29$, giving us a total gas mass of $M_g = 0.19\, M_{dm}$. In our simulations we give each SPH particle a mass between $0.25\,{\rm M_\odot}$ and $5.0\,{\rm M_\odot}$, depending on the total gas mass (for details see Table 7.1). In order to properly resolve gravitationally bound clumps (or other gravitationally bound structures) in SPH simulations, these must be represented by at least twice as many SPH particles as those included in the SPH smoothing kernel (see Section 4.2). In our simulations, our smoothing kernel encompasses approximately 40 particles for reasons of computational efficiency; the number is allowed to vary slightly, but never by more than 5 particles – and so our minimum mass resolution varies between $M_{\rm res} \simeq 80 m_{\rm part} \simeq 20\,{\rm M_\odot}$ and $400\,{\rm M_\odot}$.

To prevent artificial fragmentation or other numerical artifacts from affecting our results, it is necessary either to halt the simulation before the local Jeans mass, M_J falls below $M_{\rm res}$ in any part of the simulation volume, or to use sink particles to represent regions where $M_J < M_{\rm res}$ (see Section 4.2). We choose the latter. According to Equation 4.7 the maximum resolvable density depends on the sound speed and the total gas mass. Since the sound speed depends on the varying temperature, we have to give a lower limit for the temperature to estimate the maximum resolvable mass. We choose the temperature that corresponds to the cosmic microwave background (CMB) at the redshift of the run. Therefore, we create our sink particles at a hydrogen number density of approximately $500\,{\rm cm}^{-3}$ and with an accretion radius of approximately $0.1 r_s$.

We initialize each of our simulations with gas that is hot ($T_g = 10^4\,{\rm K}$) and fully ionized (x_e, x_{H_2}, description of fractional abundances see Section 4.5). The physical situation that these initial conditions are intended to represent is that of a protogalaxy forming within what Oh & Haiman (2003) term a "fossil" H ii region – an H ii region surrounding an ionizing source which has switched off, but the surrounding gas has not yet had time to cool and recombine. Prior to cosmological reionization, such regions should be relatively common, since the characteristic lifetime of the likely ionizing sources – massive population III stars and/or active galactic nuclei – are significantly shorter than the Hubble time.

We study two different cases. First we focus on primordial, zero-metallicity gas, and second, we study regions which had time to mix with enriched gas from supernovae explosions of the first generation of stars. For these regions we assume an average metallicity of $10^{-3} Z_\odot$. A metallicity of $Z = 10^{-3}\, Z_\odot$ is an upper limit derived from QSO absorption-line studies of the low column density Lyman-α forest at $z \sim 3$ (Pettini, 1999). Estimates of the globally-averaged metallicity produced by the sources responsible for reionization are also typically of the order of $10^{-3}\,Z_\odot$ (see, e.g. Ricotti & Ostriker, 2004). In our simulations of metal-enriched gas, we assume that mixing is efficient and that the metals are spread out uniformly throughout the computational domain. We also assume that the relative abundances of the various metals in the enriched gas are the same as in solar metallicity gas; given the wide scatter in observational determinations and theoretical predictions of abundance ratios in very low metallicity gas (see the discussion in Glover 2007), this seems to us to be the most conservative assumption. However, variations in the relative abundances of an order of magnitude or less will not significantly alter our results. We denote runs with zero metallicity with "ZM" and low metallicity with "LM".

Moreover we investigate the influence of a UV background. Already at $z = 20$ a considerable UV background may already have developed (Haiman et al., 2000; Glover & Brand, 2003). Indeed, if cosmological reionization is to occur somewhere in the redshift range $z_{\text{reion}} = 17 \pm 5$, as is suggested by the wmap polarization results (Kogut et al., 2003), there must already be a fairly strong background in place. To explore how the presence of a UV background may influence our conclusions, we have run several sets of simulations in which the strength of the UV background has been varied. These runs are named with "UV".

7.2 Zero Metallicity Gas

In this section we discuss our results for the case of primordial gas. We compare with the findings of other authors that have reported similar simulations concerning the formation of the first stars. This gives us confidence in our model and in the new conclusions we draw.

There are two evolutionary pathways that we want to separate depending on the parameters varied in the simulations: The initially ionized gas in the protogalactic halo is able either to cool sufficiently and collapse, or to remain at a relatively high temperature and not collapse towards the center of the dark matter halo. In Figure 7.1 we show a projection of the hydrogen nuclei number density on the x-y plane at two different times. The data are from model ZM25m. This model has a dark matter halo with a mass of $7.8 \times 10^5 \, M_\odot$ at a redshift of $z = 25$. We start our simulations from an equilibrium situation, therefore, the density distribution of the gas follows the density profile of the dark matter halo. In the top panel of Figure 7.1 we see a slight increase in gas density in the center of the computational volume 15 Myr after the onset of the simulations. 50 Million years later the central number density rises to approximately $1000 \, \text{cm}^{-3}$. Moving from the outer parts of the simulated region towards the center of the box, we find an increase in number density by a factor of 10^5. This is a clear indication for collapse of the gas. Nevertheless, we also have to take the temperature distribution into account to determine the dark matter halos in which stars eventually form (see Figure 7.6). In the next section we will investigate how halo mass and redshift influence the ability of zero-metallicity gas in the corresponding protogalactic halos to cool.

7.2.1 Dependence of Cooling and Collapse on Halo Mass

For the redshifts that we study here, namely $z = 15$, $z = 20$, $z = 25$ and $z = 30$, we choose three different halo masses, namely, a large, a medium and a small mass. We select the masses such that the corresponding change in ν (see Equations 3.2 and 3.3) between the different mass bins is similar at all 4 redshifts used and, in particular, the change in ν takes a value of approximately 0.5. This leads to differences in the actual value of the mass of

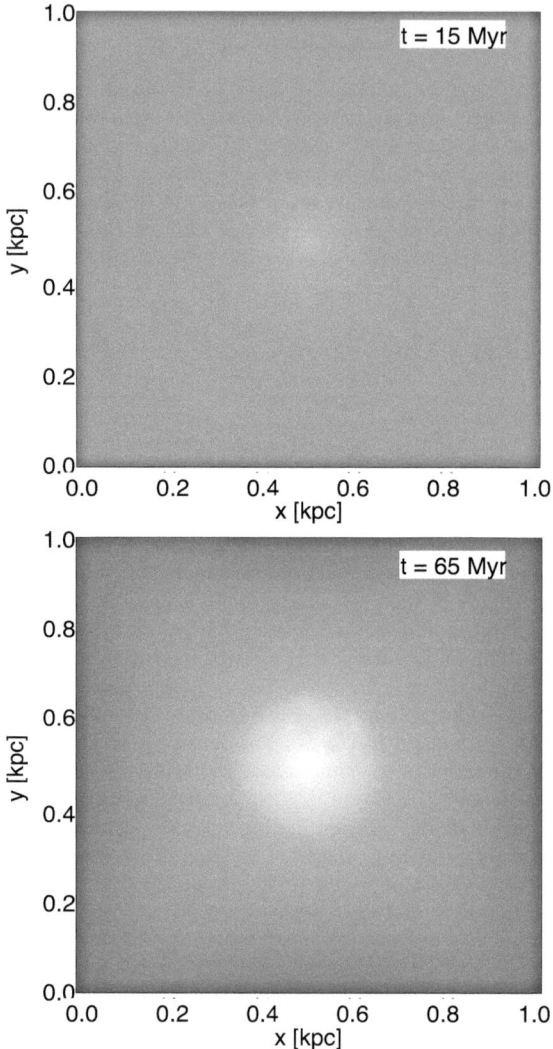

Figure 7.1: Projection of the hydrogen nuclei number density in x-y plane of run ZM25m. The density varies between $0.01\,\mathrm{cm}^{-3}$ (red) and $1000\,\mathrm{cm}^{-3}$ (white). The top panel shows the number density 15 Myr after the start of the simulation. After 65 Myr one can see the collapsed gas in the middle of the computational volume (lower panel).

the dark matter halo when comparing different redshifts. For the mass bin 'small' we take a dark matter halo with a mass above $4.0 \times 10^4 \, M_\odot$ since we do not expect efficient cooling within several Hubble times for halos less massive than this. The ν of the mass bin 'small' is different for the various redshifts since the probability of finding a dark matter halo of this mass depends on the redshift. We give the exact values of the different dark matter halo masses in Table 7.1. In Figure 7.2 we show the temporal evolution of the hydrogen nuclei number density within the scale radius r_s (see Table 7.1). The number density in this central region indicates the collapse properties of the gas in the corresponding dark matter halo. All models shown (ZM15..30s, ZM15..30m and ZM15..30l) follow gas with zero metallicity. As explained in Section 7.1 sink particles may form at number densities above $500 \, \text{cm}^{-3}$. In the top panel of Figure 7.2 we show 4 low halo mass runs at different redshifts. They all have a dark matter halo with a mass below $7.0 \times 10^4 \, M_\odot$. Only the model with redshift $z = 30$ shows an increase in central density. Nevertheless, the increase is very slow and it takes more than 400 Myrs until the gas reaches densities above $500 \, \text{cm}^{-3}$. We have to compare this time with the Hubble time at the corresponding redshift. For a redshift of 30 the Hubble time (see Equation 3.40) is approximately 80 Myr. Due to the hierachical structure formation in the universe (see Section 3.1.2) dark matter halos will probably merge within one or two Hubble times. Therefore, we are mainly interested in cooling and collapse processes that happen within times that are smaller than 2 Hubble times. The approximate Hubble times at redshifts $z = 15$, $z = 20$, $z = 25$ and $z = 30$ are $200 \, \text{Myr}$, $140 \, \text{Myr}$, $100 \, \text{Myr}$ and $80 \, \text{Myr}$, respectively (see Figure 7.2, top panel).

From the top panel of Figure 7.2 we conclude that in dark matter halos with a mass below $7.0 \times 10^4 \, M_\odot$ the zero metallicity gas does not collapse on the timescale explained above. The middle panel shows the models ZM15..30m. The protogalactic halos in these runs have masses between $5.0 \times 10^5 \, M_\odot$ and $4.2 \times 10^6 \, M_\odot$. The gas in all runs cools sufficiently to collapse in less than a Hubble time. These protogalactic halos are relativly small, especially at high redshifts, but also moderately abundant. Nevertheless the gas in these halos is able to cool and collapse, therefore, these halos are potential sites of star formation. The gas in dark matter halos of larger masses, as shown in the bottom panel of Figure 7.2, cools vary rapidly due to the increase in gravitational potential. In these halos the gas reaches higher densities where H_2 cooling is more efficient. Nevertheless, these halos have only a ν of approximately 4 at high redshifts, i.e. they are not as abundant as smaller halos. From our models we can infer that only protogalactic halos with masses above $\sim 5.0 \times 10^5 \, M_\odot$ can host cold gas. This is consistent with the result by Yoshida et al. (2003) who find cold gas clumps only in halos with masses above $\sim 7.0 \times 10^5 \, M_\odot$. In the following section we focus on one specific run to investigate the influence of the ultraviolet background on the collapse of protogalactic gas.

7.2.2 Dependence of Cooling and Collapse on UV Background

In this section we study in some detail the influence of an external UV background. As explained in Section 7.1, a considerable UV background may already have developed by $z = 20$ (Haiman et al., 2000; Glover & Brand, 2003) by reionization processes, e.g. the influence of the very first stars. We use 2 different values for J_{21} (see Equation 4.12): 0.01 and 0.1.

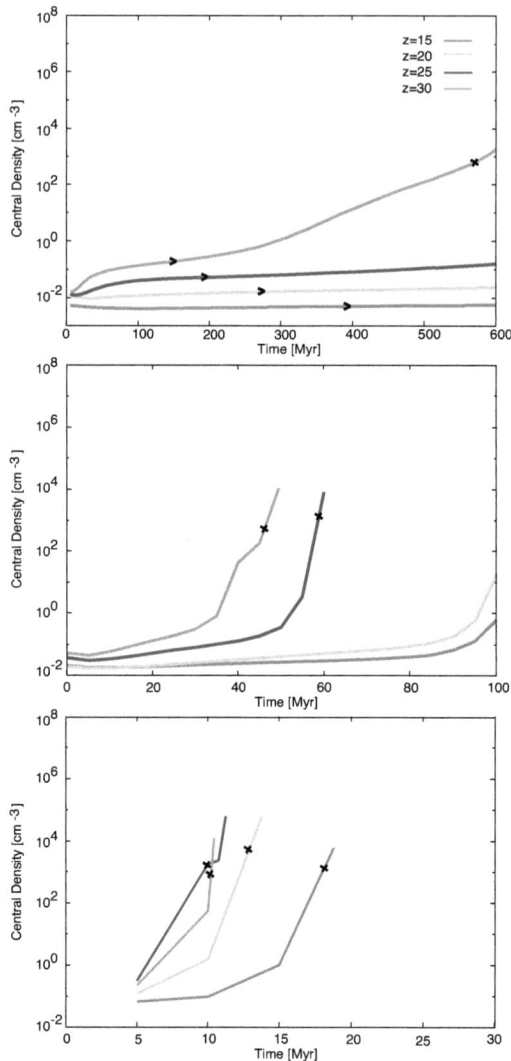

Figure 7.2: The time evolution of the number density of hydrogen nuclei within the scale radius r_s of the dark matter halo for three suites of runs in different bins of the dark matter halo mass: top panel - model ZM15..30s with small masses; middle panel - model ZM15..30m with medium masses; bottom panel - model ZM15..30l with large masses. Please note the different time axes. For clarity, we only plot the evolution up until the point at which a sink particle forms or until the end of the run, if no sink forms. The arrows in the top panel mark twice the Hubble time. The crosses show the creation of a sink particle.

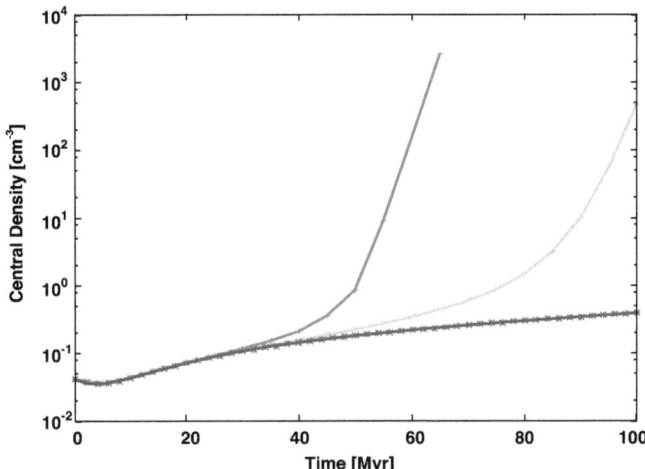

Figure 7.3: The time evolution of the gas number density within the scale radius r_s of the dark matter potential. We show runs ZM25m (plus signs), ZM25mUV (cross signs) and ZM25mUV2 (star signs) which all have zero metallicity. The UV background increases from run ZM25m to run ZM25mUV2 (see Table 7.2). For clarity, we only plot the evolution up until the point at which a sink particle forms (or until the end of the run, if no sink forms).

We also choose these values to be able to directly compare our results with other authors. In this section we concentrate for simplicity on a specific dark matter halo (model ZM25m) at a redshift of 25.

Figures 7.3 and 7.4 compare the time evolution of the number density and the temperature of the gas within the scale radius r_s for the runs with gas of zero metallicity. After 50 Myr the central gas in run ZM25m starts collapsing as explained in the previous section. In run ZM25mUV an imposed UV background of $J_{21} = 0.01$ photodissociates part of the molecular gas delaying the process of collapse and effective cooling by more than 30 Myr. A further increase in UV background by a factor of 10 inhibits the collapse of the gas within the time frame of our simulation. These results agree with the findings by Haiman et al. (2000); Machacek et al. (2001). These authors also find that UV background can delay cooling and collapse of small or medium mass halos.

Table 7.2 gives an overview on the properties of the gas within the scale radius after 100 Myr of evolution which corresponds to exactly one Hubble time. We show the minimum central temperature $T_{c,\min}$, the maximum central density $n_{c,\max}$, the maximum central fractional abundance $x_{H_2,c,\max}$ of H_2, computed with respect to the total number of hydrogen nuclei and the fraction $f_{\rm coll}$ of the initial gas mass within the virial radius of the dark matter halo that is cold and dense, i.e. with a temperature below 200 K and a density above $500\,{\rm cm}^{-3}$. In runs ZM25m and ZM25mUV a sink particle forms in the center of the dark matter halo. Since we have no information on the true temperature and density distribution

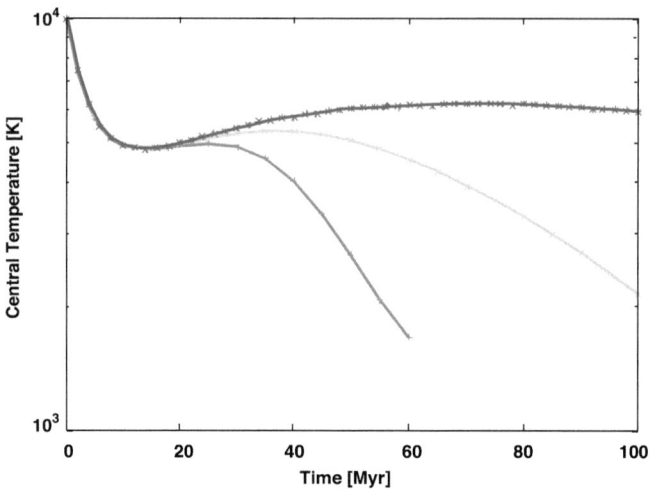

Figure 7.4: Same as Figure 7.3, but for the central temperature of the gas.

of the gas represented by the sink particle, we can only give limits. From the criterium for sink particle formation, we know the central density must exceed $n = 500\,\text{cm}^{-3}$. Similar holds for the central temperature which we know must be cooler than that of the gas just prior to sink particle formation, which corresponds to $T = 200\,\text{K}$. The high fraction of the gas within the sink particle in run ZM25m clearly shows the advanced state of the collapse which is due to the relatively high H_2 abundances achieved. In this case H_2 cooling is very effective. The UV photons that are present in the other two cases hinder the formation of H_2 by photodissociation. This decreases the maximum abundance of H_2 and delays cooling and collapse.

7.3 Low Metallicity Gas

After studying the cooling and collapse behavior of metal-free gas in protogalactic halos, in this section we turn to gas with a low metal content. After the first stars in the universe have exploded in a supernova, the formed metals have mixed with primordial gas. In addition, the gas in these halos has been fully ionized, and it is hot and metal enriched. The question that arises is how this low metal content in these H ii regions influences the cooling and collapse properties of the gas in the halo. We try to give a first answer using the chemistry and cooling routines introduced in Section 4.5.

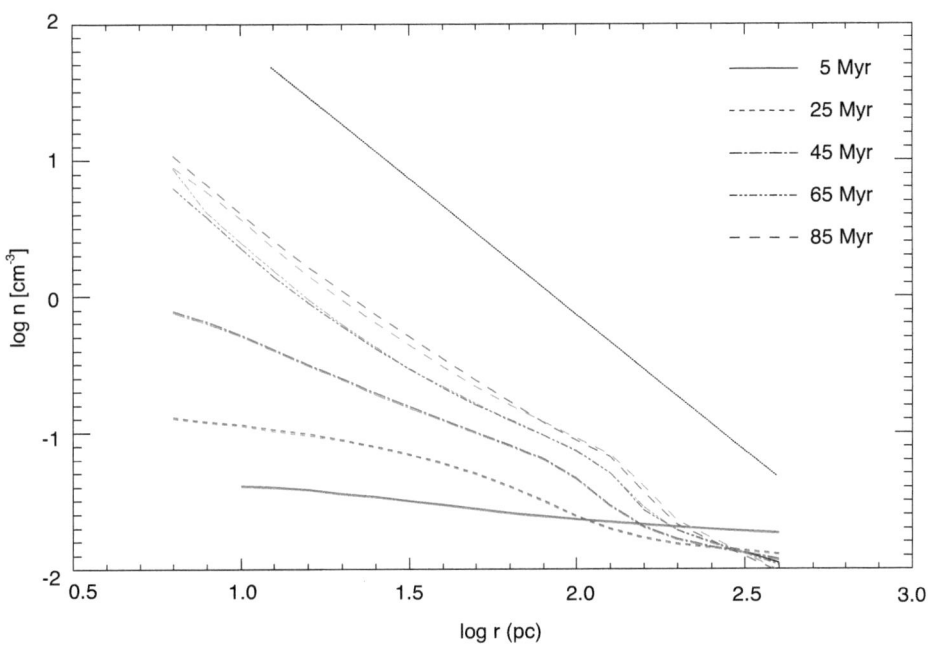

Figure 7.5: The radial dependence of the number density of the gas for models ZM25m (blue) and LM25m (red) at several times. For comparison we show a dotted line with a slope of −2.

Table 7.2: Physical state of the densest gas within the scale radius r_s after 1 Hubble time.

run	Z[a] [$Z_\odot$]	J_{21}[b]	$T_{c,min}$[c] [K]	$n_{c,max}$[d] [$Z_\odot$]	$x_{H_2,c,max}$[e] [cm^{-3}]	f_{coll}[f]
ZM25m	0.0	0.0	< 200	> 500	4×10^{-3}	0.17
LM25m	10^{-3}	0.0	< 200	> 500	2×10^{-3}	0.20
ZM25mUV	0.0	10^{-2}	< 200	> 500	5×10^{-4}	0.005
LM25mUV	10^{-3}	10^{-2}	< 200	> 500	5×10^{-4}	0.01
ZM25mUV2	0.0	10^{-1}	4400	0.7	1.6×10^{-5}	0.0
LM25mUV2	10^{-3}	10^{-1}	5900	0.6	1.5×10^{-5}	0.0

[a]Metallicity of the gas.
[b]Strength of the UV background.
[c]Minimum temperature of the gas within the scale radius r_s.
[d]Maximum number density of the gas within the scale radius r_s.
[e]Maximum fractional H$_2$ abundance within the scale radius r_s.
[f]Mass fraction of gas within a sink particle.

7.3.1 Dependence of Cooling and Collapse on Metallicity

In this section we consider whether gas with metallicity $Z = 10^{-3} Z_\odot$ changes its cooling and collapse behavior compared to the zero metallicity case. We study the influence of the metallicity for 2 specific models ZM25m and LM25m which only differ in metal content. We select these models because they represent the best cases to study the importance of small dark matter halos for the formation of the second generation of stars. In Figure 7.5 we show the temporal evolution of the radial distribution of the hydrogen nuclei number density. We directly compare the model with metal free gas to the model with gas of low metal content. The models show striking similarity in the temporal evolution. The gas collapses in both cases, and up to densities around $10\,\text{cm}^{-3}$ the evolution of the density profiles is almost indistinguishable. After 65 Myr both radial profiles approach $n \propto r^{-2}$.

From Figure 7.5 we can conclude that the metals do not influence the collapse by much. More support to this result comes from Figure 7.6. Here we show the radial profiles of temperature, fractional H$_2$ abundance and fractional H$^+$ abundance (from top to bottom) and their temporal evolution. Again, we compare the results for model ZM25m (no metallicity) on the left hand side and model LM25m on the right hand side of the figure. Before we discuss the evolution of the individual profiles, we can infer from this figure that the metal content does not result in effective changes in the radial profile of all variables. The top panels show the temporal evolution of the radial profile of the temperature during collapse. Gas that is outside the virial radius (for this example $r_{vir} \approx 100\,\text{pc}$) cools predominantly via Compton cooling and, therefore, the cooling process starts 5 Myr after initialization. For smaller radii the cooling time is much longer since cooling can only set in efficiently after enough H$_2$ has been formed. We show the evolution of the fractional H$_2$ abundance in the middle panel of Figure 7.6. Efficient cooling via H$_2$ only sets in after the fractional

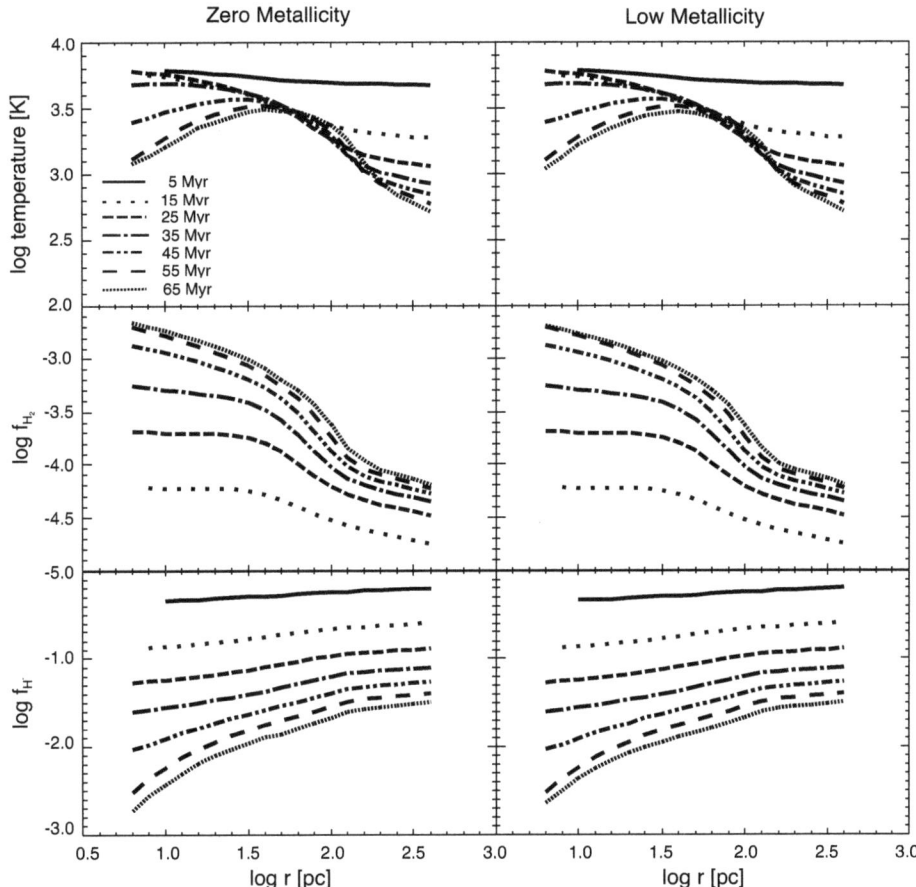

Figure 7.6: Time evolution of the radial dependence of various parameters for models ZM25m (left) and LM25m (right) at several times. We show the dependence of temperature (top panel), fractional H_2 abundance (middle panel) and fractional H^+ abundance (bottom panel).

H$_2$ abundance has reached a value of 10^{-3}. While the fractional abundance of H$_2$ rises, the fractional abundance of H$^+$ decreases. In the bottom panel of Figure 7.6 we show how the initially fully ionized gas partially recombines. This process is especially prominent in the core region where H$_2$ forms. From the above evidence we conclude that H$_2$ cooling is the dominant cooling mechanism, this is true for gas with zero metallicity as well as for gas with low metallicity. Nevertheless, it is also interesting to investigate how the most important metals O, C, Si evolve during the collapse process. The metals as well as the hydrogen are initially fully ionized. In Figure 7.7 we show the temporal evolution of the radial profiles of C ii, O ii and Si ii for the model LM25m. With time all metals recombine and the ionization fraction decreases.

7.3.2 Dependence of Cooling and Collapse on Metallicity and UV Background

The evidence in the preceding section implies that the metallicity of the gas does not dominate its cooling behavior. We arrive at this result using initially ionized gas with number densities below $10\,\mathrm{cm}^{-3}$ in small protogalactic halos. The question remains if further parameters like the UV background can change this result. We investigate this possibility in this section.

In Figures 7.8 and 7.9 we compare the time evolution of the relative difference of central gas density and central gas temperature between corresponding runs with low metallicity and runs with zero metallicity. For times below 50 Myr the difference between corresponding runs stays below 1 % of the value for zero metallicity gas. This holds true for both, the central density and the central temperature. As shown in the previous sections, the cooling of the gas is dominated by H$_2$ cooling for densities below $1\,\mathrm{cm}^{-3}$ and temperatures above 2000 K. Fine structure cooling due to the presence of metals only contributes marginally to the overall cooling of the gas. The evolution of cooling during the first 50 Myr and the onset of collapse are thus almost independent from the metallicity of the gas if the metallicity is below $Z = 10^{-3}\,Z_\odot$. Nevertheless, once the central region goes into collapse the efficiency of the metal cooling rises. Only run LM25m that collapses first shows a slight difference in the fraction of gas that is represented by a sink particle (see Table 7.2). The influence of the UV background remains the same as in the case of zero metallicity gas. Cooling and collapse is delayed or inhibited depending on the strength of the UV background. The metals in the gas cannot initiate collapse if the gas cannot go into collapse without any metals present.

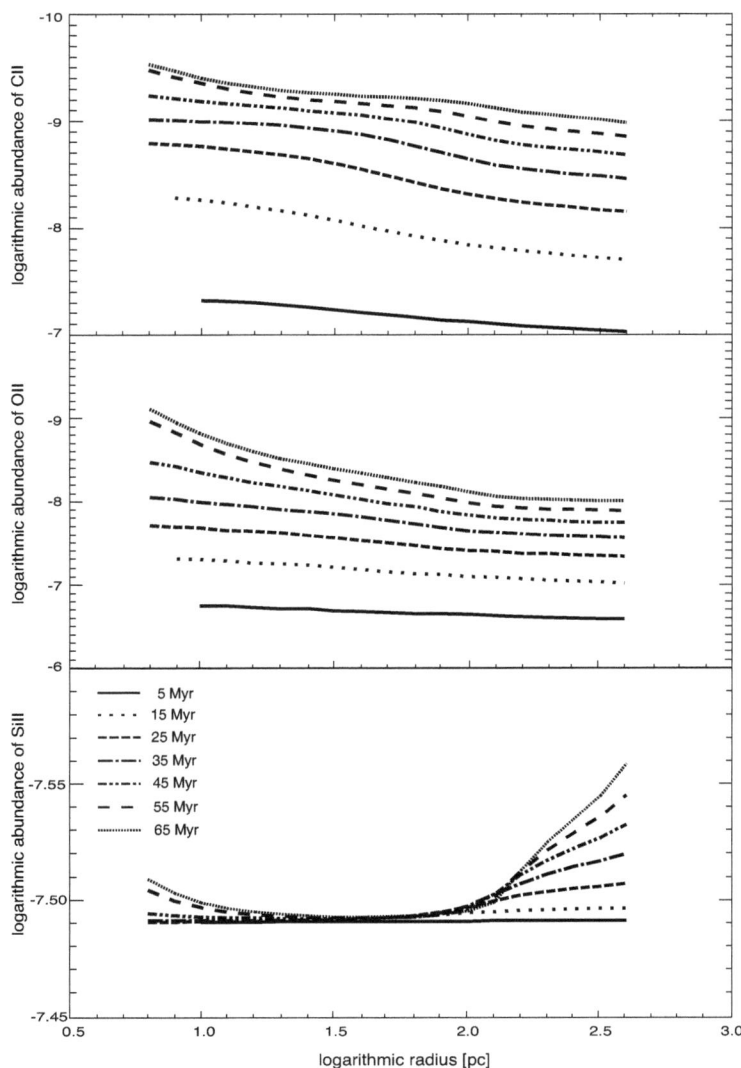

Figure 7.7: The radial dependence of various metal abundances for model LM25m at several times. We show the dependence of fractional C ii abundance (top panel), fractional O ii abundance (middle panel) and fractional Si ii abundance (bottom panel).

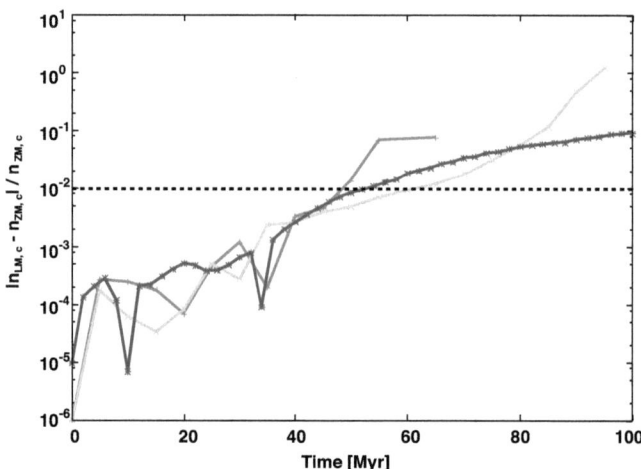

Figure 7.8: The time evolution of the relative difference $|n_{\mathrm{ZM,c}} - n_{\mathrm{LM,c}}|/n_{\mathrm{ZM,c}}$, where $n_{\mathrm{LM,c}}$ is the central gas density of the runs with low metallicity gas and $n_{\mathrm{ZM,c}}$ the central gas density in the runs with zero metallicity. We show the values for runs LM25m (plus signs), LM25mUV (cross signs) and LM25mUV2 (star signs) which all have a metallicity $Z = 10^{-3} Z_\odot$ relative to the ZM25m runs with equivalent UV background. The UV background increases from run LM25m to run LM25mUV2 (see Table 7.2). For clarity, we only plot the evolution up until the point at which a sink particle forms (or until the end of the run, if no sink forms). From Jappsen et al. (2007).

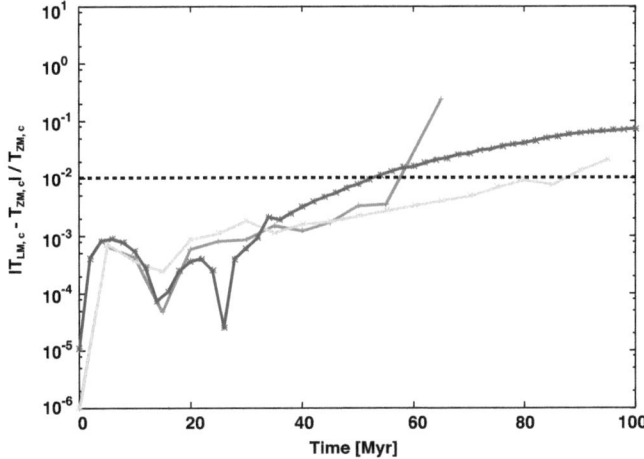

Figure 7.9: Same as Figure 7.8, but for the central temperature of the gas. From Jappsen et al. (2007).

7.4 Summary and Discussion

The results that we have presented in this chapter are a first step to a better understanding of the formation of stars in the early universe. So far many authors have modeled the dark matter halos where the first generation of stars are born. For example, calculations performed by Abel et al. (2002) show that rapid accretion rates driven by molecular hydrogen cooling cause the formation of solitary massive protostars in the range 30 to 300 $M_\odot$ in minihalos of $10^5 - 10^6 M_\odot$ at redshifts > 20. Over the main sequence lifetime of the central star (on the order $2 - 6$ Myr for the range of 30 - 300 $M_\odot$) half of the baryons within the minihalo are driven beyond its virial radius by ionized flows that quickly steepen into shocks. An important question is whether later generations of stars can efficiently form in the relatively high temperatures and ionization fractions of the relic H ii regions left by the first stars. This question is still an open issue.

Recently O'Shea et al. (2005) showed that a second primordial star can form in the relic H ii region of an earlier Pop III star. But other studies show that the feedback on the following generation of stars will be negative, i.e. prevent subsequent star formation. One analytic study (Oh & Haiman, 2003) found that the first stars injected sufficient entropy into the early intergalactic material, by photoheating and supernova explosions, to prevent further local star formation in their vicinity. UV background radiation is also thought to have contributed a negative feedback by photodissociating primordial H_2 and quenching the molecular hydrogen cooling processes that allowed the first primordial stars to form (Haiman et al., 2000; Machacek et al., 2001).

In this chapter we have adressed the issue of cooling and collapse of ionized gas in small

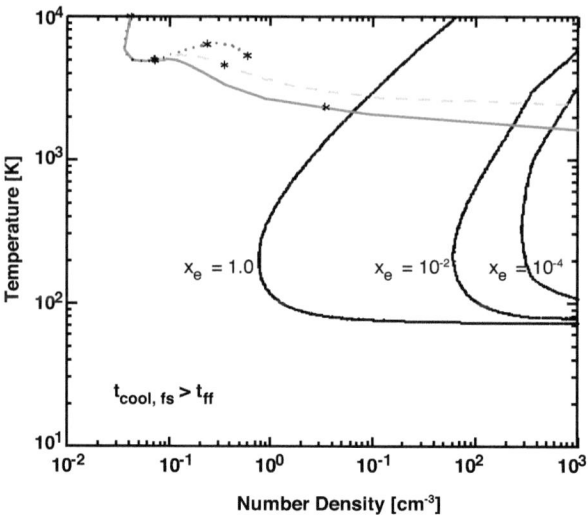

Figure 7.10: The contours labeled with x_e indicate the temperature and density at which the cooling time due to fine structure emission, $t_{\rm cool,fs}$, equals the free-fall time, $t_{\rm ff}$, for gas with a metallicity $Z = 10^{-3}\,Z_\odot$ and with fractional ionizations $x_e = 1.0$, $x_e = 10^{-2}$ and $x_e = 10^{-4}$. To the left of each line, $t_{\rm cool,fs} > t_{\rm ff}$, so metal cooling is inefficient. In every case, we assume that all of the carbon and silicon is present as C ii and Si ii respectively; in the $x_e = 1.0$ case, we assume that all of the oxygen is present in the form of O ii, but otherwise that it is all O i. To compute the free-fall time, we take the density to be the sum of the gas density $\rho_{\rm g}$ and the dark matter density at the center of the halo $\rho_{\rm dm}$. The figure also shows how the temperature and density of the gas at the centre of the halo evolve in runs L25m (solid line), L25mUV (dashed line) and L25mUV2 (dotted line). The evolution of the fractional ionization in these runs is indicated by the star symbols: $x_e = 1.0$ for $T = 10^4$ K, and decreases by a factor of 10 between each successive star. From Jappsen et al. (2007).

protogalactic halos. We find that cooling and collapse is possible even in halos with low mass ($M_{\rm DM} > 5.0 \times 10^5\,M_\odot$) and at high redshifts ($30 < z < 15$). UV background delays or quenches molecular hydrogen cooling processes. In regions with a UV background $J_{21} > 0.01$ further star formation may not be possible.

The chemical network that we have implemented in a 3-dimensional hydrodynamics code allows us to study gas with zero metallicity as well as gas with low metallicity. From our models we find that gas with low metallicity collapses very similar to gas with zero metallcity. Although the gas is pre-enriched, H_2 cooling remains the dominant process. At first sight, the fact that fine structure cooling from metals has so little impact on the thermal or dynamical evolution of the gas even when $Z = 10^{-3}\,Z_\odot$ is somewhat surprising, given that Bromm et al. (2001) found that gas with this metallicity could cool rapidly and fragment even in the absence of molecular hydrogen. However, comparison of the cooling time due to fine structure emission to the free-fall time helps to make the situation clear. The cooling time due to fine structure emission is given by

$$t_{\rm cool,fs} = \frac{1}{\gamma - 1}\frac{nkT}{\Lambda_{\rm fs}}, \tag{7.2}$$

where $\Lambda_{\rm fs}$ is the total cooling rate per unit volume due to fine structure emission, T is the temperature of the gas, γ is the adiabatic index, n is the number density of the gas and k is the Boltzmann constant. The free-fall time can be written as

$$t_{\rm ff} = \left(\frac{3\pi}{32G\rho}\right)^{1/2}. \tag{7.3}$$

where G is the gravitational constant and $\rho = \rho_{\rm gas} + \rho_{\rm dm}$ (see Equation 3.31). The variables $\rho_{\rm gas}$ and n are connected via the molecular weight μ: $\rho_{\rm gas} = \mu n$. In Figure 7.10, we indicate the temperature and density at which $t_{\rm cool,fs} = t_{\rm ff}$ for $10^{-3} Z_\odot$ gas, for three different assumed fractional ionizations, with $x_{\rm e} = 1.0$ (fully ionized gas), $x_{\rm e} = 10^{-2}$, and $x_{\rm e} = 10^{-4}$. In each case, we assume that the carbon and silicon are present only as C ii or Si ii, since a fairly small external UV flux is sufficient to achieve this. In the $x_{\rm e} = 1.0$ case, we also assume that all of the oxygen is O ii, since charge transfer between oxygen and hydrogen, which have nearly identical ionization potentials, typically ensures that $x_{\rm O^+}/x_{\rm O} \simeq x_{\rm H^+}/x_{\rm H}$. For the dark matter density $\rho_{\rm dm}$, we adopt the value found at the center of our model protogalactic halo, $\rho_{\rm dm} \simeq 8 \times 10^{-22}\,{\rm g\,cm^{-3}}$ at t=0. In the plot, regions to the left of the lines have $t_{\rm cool,fs} > t_{\rm ff}$, while those to the right have $t_{\rm cool,fs} < t_{\rm ff}$.

At the beginning of our simulations, gas in the center of the halo has a temperature $T = 10^4$ K and a number density $n = 0.03\,{\rm cm^{-3}}$. It therefore lies outside of the regime where fine structure cooling is efficient, and so it is not surprising that we find that metal line cooling is initially unimportant. As the gas cools, whether through Compton cooling or H_2 emission, and begins to compress as it falls in to the halo, it moves towards the temperature and density regime in which fine structure cooling is effective. At the same time, however, the gas is recombining, which moves the boundary of this regime to the right in the plot, towards higher densities. The physical reason for this shift is the fact that free electrons are much more effective than neutral hydrogen at exciting the C^+ and Si^+ fine structure lines, and so electron excitation dominates for $x_{\rm e} > 10^{-2}$–10^{-3}, depending on the temperature. The net effect is that fine structure cooling remains of little importance until the gas is near the high-cooling regime. This does not occur until after considerable cooling and compression has already taken place, and therefore does not occur at all if H_2 cooling is ineffective, as in runs ZM25mUV2 and LM25mUV2.

Why then do Bromm et al. (2001) come to such a different conclusion? The answer lies in the difference between the initial conditions used in their simulations and ours. They adopt a low initial temperature for the gas of 200 K at $z = 100$, and this declines further due to adiabatic cooling prior to the formation of their simulated protogalactic halo at $z \sim 30$. Since this halo has a mass of $2 \times 10^6\,M_\odot$ and a virial temperature $T_{\rm vir} \simeq 5000$ K, the gas temperature is initially very much smaller than the halo virial temperature. Consequently, gas pressure support is initially ineffective at preventing the collapse of gas into the halo. Pressure effects only become important once the gas becomes virialized, at which time it has a temperature $T_{\rm gas} = T_{\rm vir} \simeq 5000$ K, and, crucially, a characteristic number density $n \simeq 10^{2.5}$ cm^{-3}. As can be seen from Figure 7.10, gas with this combination of temperature and density lies close to or within the regime where $t_{\rm cool,fs} < t_{\rm ff}$, depending on its fractional ionization, and so it is not surprising that Bromm et al. (2001) find that fine structure cooling is effective and that the gas can cool even in the complete absence of H$_2$.

On the other hand, in our simulations the high initial temperature of 10^4 K and lower virial temperature $T_{\rm vir} \simeq 1900$K mean that $T_{\rm gas} > T_{\rm vir}$ initially, and that gas pressure support is important right from the start. Indeed, at the beginning of our simulations it precisely balances the effects of gravity, since we start with the gas in hydrostatic equilibrium. Therefore, there is no initial phase of free-fall collapse as in the Bromm et al. (2001) simulations. Instead, significant gravitational collapse occurs only if the gas is able to cool to a temperature of order $T_{\rm vir}$ or below, which, since fine structure cooling is initially ineffective, will only occur if enough H$_2$ can form in the low density gas.

Therefore, the key question is which set of initial conditions is more appropriate. It is difficult to see how intergalactic gas could become metal enriched without at some point being ionized, since previous calculations have shown that the size of a typical region enriched by a population III supernova is much smaller than the size of the H ii region created by its progenitor star (see, e.g. Bromm et al., 2003). We would thus expect our initial conditions to be more appropriate than those of Bromm et al. (2001) for treating recently enriched and ionized regions. However, if enough time elapses following the enrichment event for the gas to be able to cool down to a temperature of a few hundred K, then the Bromm et al. (2003) initial conditions will be more appropriate. As Oh & Haiman (2003) show, this is most likely to occur in high redshift gas with a low overdensity, as in this case Compton cooling is fast and highly effective and can cool the gas to $T \sim 300$ K within a recombination time. On the other hand, at lower redshifts, or at higher overdensities, the gas recombines before it can cool, and the temperature that can be reached by Compton cooling alone is much higher, as is the case in our simulated halos. Note that in either case the metals play no significant role in determining the final temperature of the gas.

An important implication of these results is that if we are primarily concerned with investigating questions such as how $M_{\rm crit}$ evolves with redshift, or how UV feedback in the form of Lyman-Werner photons affects the ability of the gas to cool, then we need not worry about the effects of metal enrichment, as the thermal evolution of the gas on the scales of interest for these questions is completely dominated by Compton cooling and/or H$_2$ cooling. Therefore, results from studies such as Haiman et al. (2000) or Yoshida et al. (2003) give a better guide to the behavior of small, low-metallicity protogalaxies than might have been anticipated (although the additional complications posed by the mechanical energy injected into the gas by H ii regions and supernovae do of course still need to be taken into account).

However, it is important to stress that our results do not address the question of whether or not there is a critical metallicity $Z_{\rm crit}$ above which fine structure cooling from metals allows efficient fragmentation to occur. This is because if fragmentation does occur, we would expect it to occur at densities $n > 500$ cm^{-3} which are unresolved in our current simulations. We intend to examine this question using much higher resolution simulations in future work.

Our results show that even with a high UV background, H_2 remains the dominant coolant up to hydrogen nuclei number densities of $n \approx 1\,\mathrm{cm}^{-3}$ irrespective of the metallicity of the gas. Nevertheless, it will be important to see if this trend continues for higher densities. We plan to access this region of the parameter space in future work by running high resolution simulations that will enable us to resolve higher densities and smaller masses. This will enable us to follow the collapse of individual protostars and thus address questions of the stellar mass function and give us more insight into the transition from massive population III stars to stars of solar metallicity.

Chapter 8

Summary and Future Prospects

Understanding the processes leading to the formation of stars is one of the fundamental challenges in astronomy. Our understanding of the fundamental physics underlying star formation has advanced greatly in recent years. Nevertheless there is still a strong demand for theoretical work to reduce the gap between observations and theory. In this thesis we study two important aspects that any theory of star formation should solve, namely the angular momentum and the mass distribution problem (Zinnecker, 2004). In particular, we concentrate on the influence of the thermodynamic properties of star-forming gas on the initial stellar mass function. However, we do not address the question of how to solve the third problem of star formation, the magnetic flux problem (only magnetically supercritical cores can collapse and form stars Appenzeller, 1982; Shu et al., 2004). This remains a task for future work.

We use 3-dimensional hydrodynamical simulations that follow the interplay between gravity, interstellar turbulence and gas pressure. From these models we find that the process of gravoturbulent fragmentation, i.e. the interplay between supersonic turbulence and the self-gravity of the gas, is able to produce many of the observed features in Galactic star forming regions. In Chapter 5 we perform a detailed analysis of the evolution of the angular momentum during collapse. With the appropriate physical scaling, we find the specific angular momentum j of prestellar cores in our models, i.e. cloud cores as yet without central protostar, to be on average $\langle j \rangle = 7 \times 10^{20}$ cm^2 s^{-1}. This agrees remarkably well with observations of cloud cores by Caselli et al. (2002) or Goodman et al. (1993). Some prestellar cores go into collapse to build up stars and stellar systems. The resulting protostellar objects have on average $\langle j \rangle = 8 \times 10^{19}$ cm^2 s^{-1}. This is one order of magnitude less, and falls into the range observed in G-dwarf binaries (Duquennoy & Mayor, 1991).

We also find that the time evolution of specific angular momentum j is intimately connected to the mass accretion history of a protostellar core. As interstellar turbulence and mutual interaction in dense clusters are highly stochastic processes, the mass growth of individual protostars is unpredictable and can be very complex. Nevertheless, in a statistical sense, we identify a clear correlation between the specific angular momentum j and mass M, that is best represented by the relation $j \propto M^{2/3}$. This can be interpreted conveniently assuming collapse of an initially uniform density sphere in solid body rotation. A collapsing cloud core can fragment further into a binary or higher-order multiple or evolve into a protostar with a stable accretion disk. It is the ratio of rotational to gravitational energy β that determines which route the object will take. The β-distribution resulting from gravoturbulent cloud fragmentation reported here agrees well with β-values derived from observations (Goodman et al., 1993). The average value is $\beta \approx 0.05$. The fact that all cores in the observational sample have $\beta < 0.2$ implies that gravitational contraction is needed to achieve density contrasts high enough for sufficiently low β. This fits in the picture of gravoturbulent fragmentation where molecular cloud structure as a whole is dominated by supersonic turbulence but stars can only form in those regions where gravity overwhelms all other forms of support. In our simulations angular momentum is lost during collapse mostly due to gravitational torques exerted by the ambient turbulent flow as well as by mutual protostellar interactions in a dense cluster environment.

In future work it will be desirable to quantify the contribution of each of these effects in more detail. Magnetic torques are not included in our models, these would lead to even larger angular momentum transport. Simulations that include self-gravity, turbulence and magnetic fields would be desirable to investigate the overall evolution of the angular momentum (see e.g. Ziegler, 2005). However, smoothed particle hydrodynamics with magnetic fields are still problematic although there have been many attempts to merge these (e.g. Maron & Howes, 2003; Price & Monaghan, 2004). Another effect which could influence the evolution of the angular momentum is feedback from evolved massive stars. Bipolar outflows, winds and the radiation from young stars are also able to deposit large amounts of energy and momentum into the surrounding molecular cloud and thus also change the angular momentum in cloud cores. The inclusion and treatment of these phenomena in the current models are the next step towards a better understanding and a more complete theory of star formation.

The second point that we discuss in this thesis is the initial stellar mass function. In particular, we focus on the dependence of the fragmentation behavior of star-forming gas on its thermodynamic state. We conclude that the detailed knowledge of the thermal properties of the gas is very important for the adequate modeling of star formation. These properties are determined by the balance between heating and cooling processes in the gas. This is not only true for star formation in the solar neighborhood but also for star formation at high redshifts. The thermodynamic properties of the gas in these environments might be considerably different but they are equally important to the corresponding process of star formation.

In Chapter 6 we approach the problem by using a piecewise polytropic equation of state (EOS) to describe the thermal evolution of the gas during collapse. This deviation from the usual isothermal description shows that changes in the EOS influence directly the characteristic mass scale for fragmentation, and consequently the peak of the initial stellar mass function. This characteristic mass can be obtained from observations and is strikingly universal in regions in the solar neighborhood. Initial conditions in these regions can vary considerably. If the IMF depends on the initial conditions, there would be thus no reason for the characteristic mass to be universal. Therefore, a derivation of the characteristic stellar mass that is based on fundamental atomic and molecular physics would be more self-consistent.

Another mass scale in star formation that depends only on fundamental physics is the opacity limit mass, a lower limit on the mass scale for fragmentation at high densities, that is determined by the onset of a high opacity to the thermal emission from dust (Low & Lynden-Bell, 1976). Nevertheless in our simulations we do not reach this mass scale.

We approximate the temperature-density relationship at lower densities with the piecewise EOS given by Larson (1985). Here we show that differences in the thermodynamic behavior of the gas may give discernible changes in the IMF. Knowledge of more detailed temperature-density relations like that of Spaans & Silk (2000, 2005) will thus add more predictive power to our models.

Nevertheless, a polytropic EOS can only serve as a first approach. In order to properly model the collapse of star-forming gas, we need to be able to follow the major chemical reactions in the gas, while also following its dynamical and thermal evolution. A common assumption is that the chemical evolution of the gas can be decoupled from its dynamical evolution, with the former never affecting the latter. Although justified in some circumstances, this assumption is not true in every case (see below). In particular, it does not allow us to study some of the most interesting problems related to star formation, such as the collapse of molecular cloud cores or the formation of stars from low metallicity gas. In recent years the advances in observational techniques as well as space-based telescopes, like the Hubble Space Telescope (HST) and the Spitzer Space Telscope, make it possible to gain more and more information about star formation processes from the solar neighborhood to distant galaxies at high redshifts. The interpretation of these observations requires appropriate models that take into account the dynamical evolution as well as the chemical properties of the gas. Our ultimate goal is the meaningful comparison between our simulations and observations of star-forming regions. In order to achieve this aim we have to be able to model the chemistry of the most important tracers, like CO, CS and NH_3 in dynamical environments. These challenges require high-resolution, 3D modeling of the collapsing region with an appropriate chemical network. For present-day metallicities these simulations are not computationally feasible even with present state-of-the-art computers and techniques, thus we focus on primordial star formation in the current study.

In Chapter 7 we take a first approach to combine a chemical network with a hydrodynamical code, in order to investigate the thermodynamic behavior of hot ionized gas of low metallicity in small protogalactic halos. These regions are the probable birthplaces of the second generation of stars in the universe. It is now clear that one of the keys to a better understanding of the early episodes of star formation in the cosmos is the detailed appreciation of the chemistry and thermodynamics of H_2. Despite the limited number of elements available, primordial gas chemistry can be very complex. Nevertheless, the amount of chemical reactions and cooling processes remain manageable within the possibilities of available computers. Therefore, the problem of the cooling and collapse of hot ionized gas in small protogalactic halos is a good starting point to test our ability to combine hydrodynamical simulations with appropriate cooling and chemistry routines.

Our simulations of regions that were ionized and heated by hard UV radiation from the first stars show that cooling and collapse of the gas to form the second generation of stars is possible if certain conditions are met. We find that the dark matter halo must have a mass above $5.0 \times 10^5 \, M_\odot$ and the UV background has to be below $J_{21} = 0.01$ for the hot, ionized gas ($T \approx 10^4$ K) to be able to cool. In this thesis we are especially interested in the influence of low metallicity on the cooling properties of the gas. Our results show that up to densities of $1 \, \text{cm}^{-3}$ and down to temperatures of 2000 K fine structure line cooling by the metals only contributes below the 1% level to the overall cooling. H_2 is still the dominant coolant. This conclusion also holds for our simulations including UV background radiation. As already discussed in Section 7.4 our results on the gas with zero metallicity confirm the findings of previous authors (Haiman et al., 2000; Machacek et al., 2001; Yoshida et al., 2003), a fact that gives us confidence in our method. These results shed new light on the simulations of Bromm et al. (2001) of pre-enriched gas. They have suggested that there is a fundamental difference between the cooling processes in gas with a metallicity of $Z = 10^{-3} Z_\odot$ and those in gas with a metallicity of $Z = 10^{-4} Z_\odot$. In their simulations these authors assumed that H_2 has been radiatively destroyed by the presence of a UV background. Also, by starting with cold gas they implicitly assumed that no extra entropy or energy had been added to the gas during its enrichment, although as Oh & Haiman (2003) have shown, this is unlikely to be the case. Our results show that even with a high UV background, H_2 remains the dominant coolant up to hydrogen nuclei number densities of $n \approx 1 \text{cm}^{-3}$. Nevertheless, it will be important to see if this trend continues at higher densities. We plan to access this region in future work by running high resolution simulations that enable us to resolve higher densities and smaller masses. This will enable us to follow the collapse of individual protostars and thus address questions of the stellar mass function, and give us more insight into the transition from massive Pop III stars to stars of solar metallicity.

Appendix A

Physical Units and Constants

Table A.1: Physical Units

Name	Symbol		Value
parsec	1 pc	=	3.085678×10^{18} cm
astronomical unit	1 AU	=	1.495979×10^{13} cm
solar mass	$1\,M_\odot$	=	1.989×10^{33} g
year	1 yr	=	3.155815×10^{7} s
solar luminosity	$1\,L_\odot$	=	3.826×10^{33} erg s^{-1}

Table A.2: Physical Constants

Name	Symbol		Value
gravitational constant	G	=	6.67259×10^{-8} cm^3 g^{-1} s^{-2}
gas constant	$\mathcal{R}$	=	8.314510×10^{7} erg K
Boltzmann constant	k_B	=	1.380658×10^{-16} erg K^{-1}
proton mass	m_p	=	1.672623×10^{-24} g
speed of light	c	=	2.998×10^{10} cm s^{-1}
Hubble constant	H	=	70 km s^{-1} Mpc^{-1}

Appendix B

Primordial and Low Metallicity Gas Chemistry

H$_2$ collisional dissociation The reaction coefficients for the collisional dissociation of H$_2$ by atomic hydrogen (reaction 10) and molecular hydrogen (reaction 11) are density dependent, since they are sensitive to the population of the vibrational and rotational levels of H$_2$. To treat the former, we use a rate coefficient

$$\log k_{\rm H} = \left(\frac{n/n_{\rm cr}}{1+n/n_{\rm cr}}\right) \log k_{\rm H,h} + \left(\frac{1}{1+n/n_{\rm cr}}\right) \log k_{\rm H,l} \qquad (\text{B.1})$$

where $k_{\rm H,l}$ is the low density limit of the collisional dissociation rate and is taken from Mac Low & Shull (1986), while $k_{\rm H,h}$ is the high density limit, taken from Lepp & Shull (1983). The critical density, $n_{\rm cr}$, is given by

$$\frac{1}{n_{\rm cr}} = \frac{x_{\rm H}}{n_{\rm cr,H}} + \frac{2x_{\rm H_2}}{n_{\rm cr,H_2}}, \qquad (\text{B.2})$$

where $n_{\rm cr,H}$ and $n_{\rm cr,H_2}$ are the critical densities in pure atomic gas with an infinitesimally dilute quantity of H$_2$ and in pure molecular gas respectively. The first of these values is taken from Lepp & Shull (1983), but has been decreased by an order of magnitude, as recommended by Martin et al. (1996); the other value comes from Shapiro & Kang (1987). To treat the collisional dissociation of H$_2$ by H$_2$ we use a similar expression

$$\log k_{\rm H_2} = \left(\frac{n/n_{\rm cr}}{1+n/n_{\rm cr}}\right) \log k_{\rm H_2,h} + \left(\frac{1}{1+n/n_{\rm cr}}\right) \log k_{\rm H_2,l} \qquad (\text{B.3})$$

where the low density limit, $k_{\rm H_2,l}$, is taken from Martin et al. (1998) and the high density limit, $k_{\rm H_2,h}$, is taken from Shapiro & Kang (1987). The collisional dissociation rates computed in this way are acceptably accurate when $n_{\rm H} \gg n_{\rm H_2}$ or $n_{\rm H} \ll n_{\rm H_2}$, but may be less accurate in gas with $n_{\rm H} \approx n_{\rm H_2}$; further study of the collisional dissociation of H$_2$ in gas which is a mixture of H and H$_2$ would be desirable to help remedy this.

Dust extinction We use a local approximation to take account of the effects of dust extinction on the photochemical rates. This approximation is similar to the one used for the

Table B.1: A list of all the gas-phase reactions included in our chemical model.

Number	Reaction	Reference
1	$H + e^- \rightarrow H^- + \gamma$	Wishart (1979)
2	$H^- + H \rightarrow H_2 + e^-$	Launay et al. (1991)
3	$H + H^+ \rightarrow H_2^+ + \gamma$	Ramaker & Peek (1976)
4	$H + H_2^+ \rightarrow H_2 + H^+$	Karpas et al. (1979)
5	$H^- + H^+ \rightarrow H + H$	Moseley et al. (1970)
6	$H^- + \gamma \rightarrow H + e^-$	Wishart (1979)
7	$H_2^+ + e^- \rightarrow H + H$	Schneider et al. (1994)
8	$H_2 + H^+ \rightarrow H_2^+ + H$	Savin et al. (2004)
9	$H_2 + e^- \rightarrow H + H + e^-$	Stibbe & Tennyson (1999)
10	$H_2 + H \rightarrow H + H + H$	See text on page 119
11	$H_2 + H_2 \rightarrow H_2 + H + H$	See text on page 119
12	$H_2 + \gamma \rightarrow H + H$	Draine & Bertoldi (1996)
13	$H + e^- \rightarrow H^+ + e^- + e^-$	Janev et al. (1987)
14	$H^+ + e^- \rightarrow H + \gamma$	Ferland et al. (1992)
15	$H^- + e^- \rightarrow H + e^- + e^-$	Janev et al. (1987)

self-shielding as described in section 4.5. We first calculate the total column density of hydrogen (in all forms) within a single smoothing length. For H_2 photodissociation, the appropriate scale factor can then be obtained directly, following Draine & Bertoldi (1996):

$$R_{\rm pd} = R_{{\rm pd},\tau=0} e^{-\tau_{d,1000}} \qquad (B.4)$$

where $R_{{\rm pd},\tau=0}$ is the photodissociation rate in the absence of dust, and where $\tau_{d,1000}$ is the optical depth due to dust at 1000Å, which is given in terms of the total hydrogen column density by

$$\tau_d = 2 \times 10^{-21} N_{\rm H,tot}. \qquad (B.5)$$

For the other photochemical rates, we use the total hydrogen column density to compute the visual extinction, A_V, using a modified form of a relationship from Draine & Bertoldi (1996):

$$A_V = 5.348 \times 10^{-22} N_{\rm H,tot} (\mathcal{D}/\mathcal{D}_\odot) \qquad (B.6)$$

where $\mathcal{D}$ is the actual dust-to-gas ratio, and $\mathcal{D}_\odot$ is the dust-to-gas ratio in solar metallicity gas. Finally, we scale the rates with A_V according to the scalings given in Le Teuff et al. (2000), which are taken in most cases from van Dishoeck (1988).

Grain surface chemistry Several of the reactions in our model involve chemistry occurring, not in the gas phase, but on the surface of interstellar dust grains. For these reactions, we use rates from the literature that were originally computed for Galactic dust, but scale the rates by $(\mathcal{D}/\mathcal{D}_\odot)$, the ratio of the actual dust-to-gas ratio to the dust-to-gas ratio in solar metallicity gas. Since we also assume that the former is directly proportional to metallicity, this essentially means that the rates scale as $(Z/Z_\odot)$.

Although there are some indications that both the composition and the size distribution of high redshift dust may differ significantly from those of Galactic dust (see, e.g. Todini & Ferrara, 2001; Nozawa et al., 2003; Schneider et al., 2004) the theoretical uncertainties remain large, and so we feel justified in continuing to use scaled Galactic rates at the present time.

Cosmic rays The ionization rate of H due to cosmic rays is a free parameter in our simulations, and is specified during the initialization of the simulation. The ionization rates for H_2, C and O are computed based on the H rate using the scalings given by Liszt (2003), while the rate for Si is assumed to be the same as that for C.

Table B.1: –Continued

Number	Reaction	Reference
16	$H^- + H \rightarrow H + H + e^-$	Janev et al. (1987)
17	$H^- + H^+ \rightarrow H_2^+ + e^-$	Poulaert et al. (1978)
18	$H_2^+ + \gamma \rightarrow H + H^+$	Dunn (1968)
19	$C^+ + e^- \rightarrow C + \gamma$	Nahar & Pradhan (1997)
20	$Si^+ + e^- \rightarrow Si + \gamma$	Nahar (2000)
21	$O^+ + e^- \rightarrow O + \gamma$	Nahar (1999)
22	$C + e^- \rightarrow C^+ + e^- + e^-$	Voronov (1997)
23	$Si + e^- \rightarrow Si^+ + e^- + e^-$	Voronov (1997)
24	$O + e^- \rightarrow O^+ + e^- + e^-$	Voronov (1997)
25	$O^+ + H \rightarrow O + H^+$	Stancil et al. (1999)
26	$O + H^+ \rightarrow O^+ + H$	Stancil et al. (1999)
27	$C + H^+ \rightarrow C^+ + H$	Stancil et al. (1998)
28	$Si + H^+ \rightarrow Si^+ + H$	Kingdon & Ferland (1996)
29	$C^+ + Si \rightarrow C + Si^+$	Le Teuff et al. (2000)
30	$C + \gamma \rightarrow C^+ + e^-$	Verner et al. (1996)
31	$Si + \gamma \rightarrow Si^+ + e^-$	Verner et al. (1996)

Table B.1: –Continued

Number	Reaction	Reference
32	$H + c.r. \rightarrow H^+ + e^-$	Liszt (2003)
33	$H_2 + c.r. \rightarrow H_2^+ + e^-$	Liszt (2003)
34	$C + c.r. \rightarrow C^+ + e^-$	Liszt (2003)
35	$O + c.r. \rightarrow O^+ + e^-$	Liszt (2003)
36	$Si + c.r. \rightarrow Si^+ + e^-$	—

Adapted from Glover (2005). References are to the primary source of data for each reaction. Photochemical reactions assume an incident spectrum corresponding to a modified, diluted black-body, as described in the text.

Table B.2: A list of all the grain surface reactions included in our chemical model.

Number	Reaction	Reference
S1	$H + H \rightarrow H_2$	Hollenbach & McKee (1979)
S2	$H^+ + e^- \rightarrow H$	Weingartner & Draine (2001)
S3	$C^+ + e^- \rightarrow C$	Weingartner & Draine (2001)
S4	$Si^+ + e^- \rightarrow Si$	Weingartner & Draine (2001)

References are to the primary source of data for each reaction. A grain size distribution and composition appropriate to Milky Way dust is assumed.

Bibliography

Abel, T., Anninos, P., Zhang, Y. & Norman, M.L. (1997). Modeling Primordial Gas in Numerical Cosmology. New Astronomy, 2, 181–207.

Abel, T., Bryan, G.L. & Norman, M.L. (2002). The Formation of the First Star in the Universe. Science, 295, 93–98.

Adams, F.C. & Fatuzzo, M. (1996). A Theory of the Initial Mass Function for Star Formation in Molecular Clouds. ApJ, 464, 256.

Adams, T.F. (1976). The Detectability of Deuterium Lyman Alpha in QSOs. A&A, 50, 461.

Appenzeller, I. (1982). Star formation and pre-main-sequence stellar evolution. Fundamentals of Cosmic Physics, 7, 313–362.

Arcoragi, J., Bonnell, I., Martel, H., Bastien, P. & Benz, W. (1991). Fragmentation of elongated cylindrical clouds. II - Polytropic clouds. ApJ, 380, 476–483.

Ballesteros-Paredes, J., Hartmann, L. & Vázquez-Semadeni, E. (1999a). Turbulent Flow-driven Molecular Cloud Formation: A Solution to the Post-T Tauri Problem? ApJ, 527, 285–297.

Ballesteros-Paredes, J., Vázquez-Semadeni, E. & Scalo, J. (1999b). Clouds as Turbulent Density Fluctuations: Implications for Pressure Confinement and Spectral Line Data Interpretation. ApJ, 515, 286–303.

Ballesteros-Paredes, J., Klessen, R.S. & Vázquez-Semadeni, E. (2003). Dynamic Cores in Hydrostatic Disguise. ApJ, 592, 188–202.

Ballesteros-Paredes, J., Gazol, A., Kim, J., Klessen, R.S., Jappsen, A.K. & Tejero, E. (2006). The Mass Spectra of Cores in Turbulent Molecular Clouds and Implications for the Initial Mass Function. ApJ, 637, 384–391.

Barnes, J. & Hut, P. (1986). A Hierarchical O(NlogN) Force-Calculation Algorithm. Nature, 324, 446–449.

Barrado y Navascués, D., Stauffer, J.R., Bouvier, J. & Martín, E.L. (2001). From the Top to the Bottom of the Main Sequence: A Complete Mass Function of the Young Open Cluster M35. ApJ, 546, 1006–1018.

Barranco, J.A. & Goodman, A.A. (1998). Coherent Dense Cores. I. NH3 Observations. ApJ, 504, 207.

Basu, S. & Mouschovias, T.C. (1994). Magnetic braking, ambipolar diffusion, and the formation of cloud cores and protostars. 1: Axisymmetric solutions. ApJ, 432, 720–741.

Basu, S. & Mouschovias, T.C. (1995). Magnetic Braking, Ambipolar Diffusion, and the Formation of Cloud Cores and Protostars. II. A Parameter Study. ApJ, 452, 386.

Bate, M.R. & Burkert, A. (1997). Resolution requirements for smoothed particle hydrodynamics calculations with self-gravity. MNRAS, 288, 1060–1072.

Bate, M.R., Bonnell, I.A. & Price, N.M. (1995). Modelling Accretion in Protobinary Systems. MNRAS, 277, 362–376.

Becker et al. (2001). Evidence for Reionization at z~6: Detection of a Gunn-Peterson Trough in a z=6.28 Quasar. AJ, 122, 2850–2857.

Bennett et al. (2003). First-Year Wilkinson Microwave Anisotropy Probe (WMAP) Observations: Preliminary Maps and Basic Results. ApJS, 148, 1–27.

Benson, A.J., Frenk, C.S., Baugh, C.M., Cole, S. & Lacey, C.G. (2001). The Clustering Evolution of the Galaxy Distribution. MNRAS, 327, 1041–1056.

Benson, P.J. & Myers, P.C. (1989). A survey for dense cores in dark clouds. ApJS, 71, 89–108.

Benz, W. (1990). Smooth Particle Hydrodynamics - a Review. In Numerical Modelling of Nonlinear Stellar Pulsations Problems and Prospects, ed. J. R. Buchler (Dordrecht: Kluwer), 269.

Binney, J. & Tremaine, S. (1987). Galactic dynamics. (Princeton University Press, Princeton).

Black, J.H. & Dalgarno, A. (1977). Models of interstellar clouds. I - The Zeta Ophiuchi cloud. ApJS, 34, 405–423.

Blitz, L. (1993). Giant molecular clouds. In Protostars and Planets III, ed. by E.H. Levy and J.I Lunine, (University of Arizona, Tucson), 125–161.

Bodenheimer, P. (1995). Angular Momentum Evolution of Young Stars and Disks. ARA&A, 33, 199–238.

Bodenheimer, P., Burkert, A., Klein, R.I. & Boss, A.P. (2000). Multiple Fragmentation of Protostars. Protostars and Planets IV, ed. V. Mannings, A. P. Boss & S. S. Russell (Tucson: Univ. of Arizona Press), 675.

Boland, W. & de Jong, T. (1984). Hydrostatic Models of Molecular Clouds. II - Steady State Models of Spherical Clouds. A&A, 134, 87–98.

Boldyrev, S. (2002). Kolmogorov-Burgers Model for Star-forming Turbulence. ApJ, 569, 841–845.

Boldyrev, S., Nordlund, Å. & Padoan, P. (2002a). Scaling Relations of Supersonic Turbulence in Star-forming Molecular Clouds. ApJ, 573, 678–684.

Boldyrev, S., Nordlund, Å. & Padoan, P. (2002b). Supersonic Turbulence and Structure of Interstellar Molecular Clouds. Phys. Rev. Lett., 89, 31102.

Bonazzola, S., Perault, M., Puget, J.L., Heyvaerts, J., Falgarone, E. & Panis, J.F. (1992). Jeans collapse of turbulent gas clouds - Tentative theory. Journal of Fluid Mechanics, 245, 1–28.

Bond, J.R., Cole, S., Efstathiou, G. & Kaiser, N. (1991). Excursion Set Mass Functions for Hierarchical Gaussian Fluctuations. ApJ, 379, 440–460.

Bonnell, I.A. (1994). A New Binary Formation Mechanism. MNRAS, 269, 837–848.

Bonnell, I.A., Bate, M.R., Clarke, C.J. & Pringle, J.E. (2001a). Competitive Accretion in Embedded Stellar Clusters. MNRAS, 323, 785–794.

Bonnell, I.A., Clarke, C.J., Bate, M.R. & Pringle, J.E. (2001b). Accretion in Stellar Clusters and the Initial Mass Function. MNRAS, 324, 573–579.

Bonnor, W.B. (1956). Boyle's Law and Gravitational Instability. MNRAS, 116, 351.

Bonnor, W.B. (1957). Jeans' Formula for Gravitational Instability. MNRAS, 117, 104.

Boss, A.P. (1993). Collapse and Fragmentation of Molecular Cloud Cores. I - Moderately centrally condensed cores. ApJ, 410, 157–167.

Boss, A.P. (1999). Collapse and Fragmentation of Molecular Cloud Cores. VI. Slowly Rotating Magnetic Clouds. ApJ, 520, 744–750.

Boss, A.P. & Bodenheimer, P. (1979). Fragmentation in a Rotating Protostar - A Comparison of Two Three-dimensional Computer Codes. ApJ, 234, 289–295.

Bourke, T.L., Myers, P.C., Robinson, G. & Hyland, A.R. (2001). New OH Zeeman Measurements of Magnetic Field Strengths in Molecular Clouds. ApJ, 554, 916–932.

Bromm, V. & Larson, R.B. (2004). The First Stars. ARA&A, 42, 79–118.

Bromm, V., Ferrara, A., Coppi, P.S. & Larson, R.B. (2001). The Fragmentation of Pre-enriched Primordial Objects. MNRAS, 328, 969–976.

Bromm, V., Coppi, P.S. & Larson, R.B. (2002). The Formation of the First Stars. I. The Primordial Star-forming Cloud. ApJ, 564, 23–51.

Bromm, V., Yoshida, N. & Hernquist, L. (2003). The First Supernova Explosions in the Universe. ApJL, 596, L135–L138.

Brown, V., Byrne, G.D. & Hindmarsh, C., A (1989). VODE: A Variable Step ODE Solver. SIAM J. Sci. Stat. Comput., 10, 1038–1051.

Brunt, C.M., Heyer, M.H., Zivkov, V. & Mac Low, M.M. (2005). Modification of Projected Velocity Power Spectra by Density Inhomogeneities in Compressible Supersonic Turbulence. ApJ(L), in revision.

Burkert, A. & Bodenheimer, P. (2000). Turbulent Molecular Cloud Cores: Rotational Properties. ApJ, 543, 822–830.

Burles, S., Nollett, K.M. & Turner, M.S. (2001). Big Bang Nucleosynthesis Predictions for Precision Cosmology. ApJL, 552, L1–L5.

Burton, M.G., Hollenbach, D.J. & Tielens, A.G.G.M. (1990). Line Emission from Clumpy Photodissociation Regions. ApJ, 365, 620–639.

Carroll, S.M., Press, W.H. & Turner, E.L. (1992). The Cosmological Constant. ARA&A, 30, 499–542.

Caselli, P., Benson, P.J., Myers, P.C. & Tafalla, M. (2002). Dense Cores in Dark Clouds. XIV. N_2H^+ (1-0) Maps of Dense Cloud Cores. ApJ, 572, 238–263.

Cen, R. (1992). A Hydrodynamic Approach to Cosmology - Methodology. ApJS, 78, 341–364.

Cernicharo, J. (1991). The Physical Conditions of Low Mass Star Forming Regions. In NATO ASIC Proc. 342: The Physics of Star Formation and Early Stellar Evolution, 287.

Chabrier, G. (2003). Galactic Stellar and Substellar Initial Mass Function. PASP, 115, 763–795.

Chandrasekhar, S. (1949). Turbulence - a Physical Theory of Astrophysical Interest. ApJ, 110, 329.

Chandrasekhar, S. (1951a). Proc. R. Soc. London A, 210, 26.

Chandrasekhar, S. (1951b). The Fluctuation of Density in Isotropic Turbulence. Proc. R. Soc. London A, 210, 18.

Chandrasekhar, S. (1967). An Introduction to the Study of Stellar Structure. (Dover Publications, New York).

Chandrasekhar, S. & Fermi, E. (1953a). Magnetic Fields in Spiral Arms. ApJ, 118, 113.

Chandrasekhar, S. & Fermi, E. (1953b). Problems of Gravitational Stability in the Presence of a Magnetic Field. ApJ, 118, 116.

Chappell, D. & Scalo, J. (2001). Multifractal Scaling, Geometrical Diversity, and Hierarchical Structure in the Cool Interstellar Medium. ApJ, 551, 712–729.

Ciardi, B. & Ferrara, A. (2005). The First Cosmic Structures and Their Effects. Space Science Reviews, 116, 625–705.

Cojazzi, P., Bressan, A., Lucchin, F., Pantano, O. & Chavez, M. (2000). Zero-metallicity Stellar Sources and the Reionization Epoch. MNRAS, 315, L51–L55.

Cooray, A. & Sheth, R. (2002). Halo models of large scale structure. Phys. Rep., 372, 1–129.

Crutcher, R., Heiles, C. & Troland, T. (2003). Observations of Interstellar Magnetic Fields. LNP Vol. 614: Turbulence and Magnetic Fields in Astrophysics, 614, 155–181.

Crutcher, R.M. (1999). Magnetic Fields in Molecular Clouds: Observations Confront Theory. ApJ, 520, 706–713.

Curry, C.L. (2002). Shapes of Molecular Cloud Cores and the Filamentary Mode of Star Formation. ApJ, 576, 849–859.

de Avillez, M.A. & Mac Low, M.M. (2002). Mixing Timescales in a Supernova-driven Interstellar Medium. ApJ, 581, 1047–1060.

Draine, B.T. & Bertoldi, F. (1996). Structure of Stationary Photodissociation Fronts. ApJ, 468, 269.

Dunn, G.H. (1968). Photodissociation of H_2^+ and D_2^+: Theory. Physical Review, 172, 1–7.

Duquennoy, A. & Mayor, M. (1991). Multiplicity among Solar-type Stars in the Solar Neighbourhood. II - Distribution of the Orbital Elements in an Unbiased Sample. A&A, 248, 485–524.

Durisen, R.H., Sterzik, M.F. & Pickett, B.K. (2001). A Two-step Initial Mass Function:. Consequences of Clustered Star Formation for Binary Properties. A&A, 371, 952–962.

Ebert, R. (1957). Zur Instabilität kugelsymmetrischer Gasverteilungen. Zeitschrift für Astrophysik, 42, 263.

Ebert, R., von Hörner, S. & Temesváry, S. (1960). Die Entstehung von Sternen durch Kondensation diffuser Materie. (Springer-Verlag, Berlin).

Elmegreen, B.G. (1997a). Cloud/Intercloud Structure from Nonlinear Magnetic Waves. ApJ, 480, 674.

Elmegreen, B.G. (1997b). The Initial Stellar Mass Function from Random Sampling in a Turbulent Fractal Cloud. ApJ, 486, 944.

Elmegreen, B.G. (1999). The Stellar Initial Mass Function from Random Sampling in Hierarchical Clouds. II. Statistical Fluctuations and a Mass Dependence for Starbirth Positions and Times. ApJ, 515, 323–336.

Elmegreen, B.G. (2000a). Modeling a High-Mass Turn-Down in the Stellar Initial Mass Function. ApJ, 539, 342–351.

Elmegreen, B.G. (2000b). Star Formation in a Crossing Time. ApJ, 530, 277–281.

Elmegreen, B.G. (2000c). Two Stellar Mass Functions Combined into one by the Random Sampling Model of the Initial Mass Function. MNRAS, 311, L5–L8.

Elmegreen, B.G. (2002). A Fractal Origin for the Mass Spectrum of Interstellar Clouds. II. Cloud Models and Power-Law Slopes. ApJ, 564, 773–781.

Elmegreen, B.G. & Mathieu, R.D. (1983). Monte Carlo Simulations of the Initial Stellar Mass Function. MNRAS, 203, 305–315.

Elmegreen, B.G. & Scalo, J. (2004). Interstellar Turbulence I: Observations and Processes. ARA&A, 42, 211–273.

Elmegreen, B.G. & Scalo, J. (2006). The Effect of Star Formation History on the Inferred Stellar Initial Mass Function. ApJ, 636, 149–157.

Evans, N.J. (1999). Physical Conditions in Regions of Star Formation. ARA&A, 37, 311–362.

Evans, N.J., Rawlings, J.M.C., Shirley, Y.L. & Mundy, L.G. (2001). Tracing the Mass during Low-Mass Star Formation. II. Modeling the Submillimeter Emission from Preprotostellar Cores. ApJ, 557, 193–208.

Falgarone, E. & Phillips, T.G. (1990). A signature of the intermittency of interstellar turbulence - The wings of molecular line profiles. ApJ, 359, 344–354.

Falgarone, E., Panis, J.F., Heithausen, A., Perault, M., Stutzki, J., Puget, J.L. & Bensch, F. (1998). The IRAM key-project: small-scale structure of pre-star-forming regions. I. Observational results. A&A, 331, 669–696.

Ferland, G.J., Peterson, B.M., Horne, K., Welsh, W.F. & Nahar, S.N. (1992). Anisotropic Line Emission and the Geometry of the Broad-line Region in Active Galactic Nuclei. ApJ, 387, 95–108.

Field, G.B., Goldsmith, D.W. & Habing, H.J. (1969). Cosmic-Ray Heating of the Interstellar Gas. ApJL, 155, L149.

Fisher, R.T. (2004). A Turbulent Interstellar Medium Origin of the Binary Period Distribution. ApJ, 600, 769–780.

Fixsen, D.J., Cheng, E.S., Gales, J.M., Mather, J.C., Shafer, R.A. & Wright, E.L. (1996). The Cosmic Microwave Background Spectrum from the Full COBE FIRAS Data Set. ApJ, 473, 576.

Fleck, R.C. (1982). Star Formation in Turbulent Molecular Clouds - The Initial Stellar Mass Function. MNRAS, 201, 551–559.

Flower, D.R. & Pineau des Forêts, G. (2001). The Thermal Balance of the First Structures in the Primordial Gas. MNRAS, 323, 672–676.

Froebrich, D. & Scholz, A. (2003). Young Stars and Outflows in the Globule IC 1396 W. A&A, 407, 207–212.

Galli, D. & Palla, F. (1998). The Chemistry of the Early Universe. A&A, 335, 403–420.

Galli, D., Shu, F.H., Laughlin, G. & Lizano, S. (2001). Singular Isothermal Disks. II. Nonaxisymmetric Bifurcations and Equilibria. ApJ, 551, 367–386.

Gammie, C.F. & Ostriker, E.C. (1996). Can Nonlinear Hydromagnetic Waves Support a Self-gravitating Cloud? ApJ, 466, 814.

Gammie, C.F., Lin, Y., Stone, J.M. & Ostriker, E.C. (2003). Analysis of Clumps in Molecular Cloud Models: Mass Spectrum, Shapes, Alignment, and Rotation. ApJ, 592, 203–216.

Genzel, R. (1991). Physical Conditions and Heating/Cooling Processes in High Mass Star Formation Regions. In The Physics of Star Formation and Early Stellar Evolution, ed. by C.J. Lada and N.D. Kylafis,(Kluwer, Dordrecht), 155.

Gingold, R.A. & Monaghan, J.J. (1977). Smoothed Particle Hydrodynamics - Theory and Application to Non-spherical Stars. MNRAS, 181, 375–389.

Glover, S. (2005). The Formation Of The First Stars In The Universe. Space Science Reviews, 117, 445–508.

Glover, S.C.O. (2003). Comparing Gas-Phase and Grain-catalyzed H_2 Formation. ApJ, 584, 331–338.

Glover, S.C.O. (2007). Radiative feedback from ionized gas. MNRAS, 379, 1352–1358.

Glover, S.C.O. & Brand, P.W.J.L. (2003). Radiative Feedback from an Early X-ray Background. MNRAS, 340, 210–226.

Goldsmith, P.F. (2001). Molecular Depletion and Thermal Balance in Dark Cloud Cores. ApJ, 557, 736–746.

Goldsmith, P.F. & Langer, W.D. (1978). Molecular cooling and thermal balance of dense interstellar clouds. ApJ, 222, 881–895.

Goodman, A.A., Benson, P.J., Fuller, G.A. & Myers, P.C. (1993). Dense Cores in Dark Clouds. VIII - Velocity Gradients. ApJ, 406, 528–547.

Goodman, A.A., Jones, T.J., Lada, E.A. & Myers, P.C. (1995). Does Near-Infrared Polarimetry Reveal the Magnetic Field in Cold Dark Clouds? ApJ, 448, 748.

Gray, M.E., Taylor, A.N., Meisenheimer, K., Dye, S., Wolf, C. & Thommes, E. (2002). Probing the Distribution of Dark Matter in the A901/902 Supercluster with Weak Lensing. ApJ, 568, 141–162.

Haiman, Z., Abel, T. & Rees, M.J. (2000). The Radiative Feedback of the First Cosmological Objects. ApJ, 534, 11–24.

Halbwachs, J.L., Mayor, M., Udry, S. & Arenou, F. (2003). Multiplicity among solar-type stars. III. Statistical properties of the F7-K binaries with periods up to 10 years. A&A, 397, 159–175.

Hambly, N.C., Hodgkin, S.T., Cossburn, M.R. & Jameson, R.F. (1999). Brown Dwarfs in the Pleiades and the Initial Mass Function Across the Stellar/Substellar Boundary. MNRAS, 303, 835–844.

Hartmann, L. (2002). Flows, Fragmentation, and Star Formation. I. Low-Mass Stars in Taurus. ApJ, 578, 914–924.

Hayashi, C. (1966). Evolution of Protostars. ARA&A, 4, 171.

Hayashi, C. & Nakano, T. (1965). Thermal and Dynamical Properties of a Protostar and Its Contraction to the Stage of Quasi-Static Equilibrium. Prog. Theor. Phys., 34, 754.

Heitsch, F., Mac Low, M.M. & Klessen, R.S. (2001). Gravitational Collapse in Turbulent Molecular Clouds. II. Magnetohydrodynamical Turbulence. ApJ, 547, 280–291.

Henry, J.P., Gioia, I.M., Maccacaro, T., Morris, S.L., Stocke, J.T. & Wolter, A. (1992). The Extended Medium Sensitivity Survey Distant Cluster Sample - X-ray Data and Interpretation of the Luminosity Evolution. ApJ, 386, 408–419.

Hernquist, L., Bouchet, F.R. & Suto, Y. (1991). Application of the Ewald method to cosmological N-body simulations. ApJS, 75, 231–240.

Hillenbrand, L.A. (1997). On the Stellar Population and Star-Forming History of the Orion Nebula Cluster. AJ, 113, 1733–1768.

Hillenbrand, L.A. & Carpenter, J.M. (2000). Constraints on the Stellar/Substellar Mass Function in the Inner Orion Nebula Cluster. ApJ, 540, 236–254.

Hillenbrand, L.A. & Hartmann, L.W. (1998). A Preliminary Study of the Orion Nebula Cluster Structure and Dynamics. ApJ, 492, 540.

Hollenbach, D. & McKee, C.F. (1979). Molecule Formation and Infrared Emission in fast Interstellar Shocks. I Physical Processes. ApJS, 41, 555–592.

Hubble, E. (1929). A Relation between Distance and Radial Velocity among Extra-Galactic Nebulae. Proceedings of the National Academy of Science, 15, 168–173.

Janev, R.K., Langer, W.D. & Evans, K. (1987). Elementary Processes in Hydrogen-Helium Plasmas - Cross Sections and Reaction Rate Coefficients. Springer Series on Atoms and Plasmas, (Springer, Berlin).

Jappsen, A.K. & Klessen, R.S. (2004). Protostellar Angular Momentum Evolution during Gravoturbulent Fragmentation. A&A, 423, 1–12.

Jappsen, A.K., Klessen, R.S., Larson, R.B., Li, Y. & Mac Low, M.M. (2005). The stellar mass spectrum from non-isothermal gravoturbulent fragmentation. A&A, 435, 611–623.

Jappsen, A.K., Glover, S.C.O., Klessen, R.S. & Mac Low, M.M. (2007). Star Formation at Very Low Metallicity. II. On the Insignificance of Metal-Line Cooling During the Early Stages of Gravitational Collapse. ApJ, 660, 1332–1343.

Jeans, J.H. (1902). . Philos. Trans. R. Soc. London, A, 199, 1.

Jijina, J., Myers, P.C. & Adams, F.C. (1999). Dense Cores Mapped in Ammonia: A Database. ApJS, 125, 161–236.

Kamazaki, T., Saito, M., Hirano, N., Umemoto, T. & Kawabe, R. (2003). Molecular Outflow Search in the ρ Ophiuchi A and B2 Regions. ApJ, 584, 357–367.

Karpas, Z., Anicich, V. & Huntress, W.T. (1979). An Ion Cyclotron Resonance Study of Reactions of Ions with Hydrogen Atoms. J. Chem. Phys., 70, 2877–2881.

Kawachi, T. & Hanawa, T. (1998). Gravitational Collapse of Filamentary Clouds. PASJ, 50, 577–586.

Kim, W. & Ostriker, E.C. (2001). Amplification, Saturation, and Q Thresholds for Runaway: Growth of Self-Gravitating Structures in Models of Magnetized Galactic Gas Disks. ApJ, 559, 70–95.

Kingdon, J.B. & Ferland, G.J. (1996). Rate Coefficients for Charge Transfer between Hydrogen and the First 30 Elements. ApJS, 106, 205.

Klessen, R.S. (2001). The Formation of Stellar Clusters: Mass Spectra from Turbulent Molecular Cloud Fragmentation. ApJ, 556, 837–846.

Klessen, R.S. & Burkert, A. (2000). The Formation of Stellar Clusters: Gaussian Cloud Conditions. I. ApJS, 128, 287–319.

Klessen, R.S. & Burkert, A. (2001). The Formation of Stellar Clusters: Gaussian Cloud Conditions. II. ApJ, 549, 386–401.

Klessen, R.S. & Lin, D.N. (2003). Diffusion in supersonic turbulent compressible flows. Phys. Rev. E, 67, 046311.

Klessen, R.S., Burkert, A. & Bate, M.R. (1998). Fragmentation of Molecular Clouds: The Initial Phase of a Stellar Cluster. ApJL, 501, L205.

Klessen, R.S., Heitsch, F. & Mac Low, M.M. (2000). Gravitational Collapse in Turbulent Molecular Clouds. I. Gasdynamical Turbulence. ApJ, 535, 887–906.

Klessen, R.S., Ballesteros-Paredes, J., Vázquez-Semadeni, E. & Durán-Rojas, C. (2005). Quiescent and Coherent Cores from Gravoturbulent Fragmentation. ApJ, 620, 786–794.

Klessen, R.S., Spaans, M. & Jappsen, A.K. (2007). The stellar mass spectrum in warm and dusty gas: deviations from Salpeter in the Galactic centre and in circumnuclear starburst regions. MNRAS, 374, L29–L33.

Kogut, A., Spergel, D.N., Barnes, C., Bennett, C.L., Halpern, M., Hinshaw, G., Jarosik, N., Limon, M., Meyer, S.S., Page, L., Tucker, G.S., Wollack, E. & Wright, E.L. (2003). First-Year Wilkinson Microwave Anisotropy Probe (WMAP) Observations: Temperature-Polarization Correlation. ApJS, 148, 161–173.

Kolmogorov, A.N. (1941). The local structure of turbulence in incompressible viscous fluid for very large Reynolds numbers. Dokl. Akad. Nauk SSSR, 30, 301–305.

Koyama, H. & Inutsuka, S. (2000). Molecular Cloud Formation in Shock-compressed Layers. ApJ, 532, 980–993.

Kroupa, P. (1995a). Inverse dynamical population synthesis and star formation. MNRAS, 277, 1491.

Kroupa, P. (1995b). The dynamical properties of stellar systems in the Galactic disc. MNRAS, 277, 1507.

Kroupa, P. (2001). On the variation of the initial mass function. MNRAS, 322, 231–246.

Kroupa, P. (2002). The Initial Mass Function of Stars: Evidence for Uniformity in Variable Systems. Science, 295, 82–91.

Kroupa, P., Tout, C.A. & Gilmore, G. (1990). The low-luminosity stellar mass function. MNRAS, 244, 76–85.

Kroupa, P., Tout, C.A. & Gilmore, G. (1993). The distribution of low-mass stars in the Galactic disc. MNRAS, 262, 545–587.

Langer, W.D., van Dishoeck, E.F., Bergin, E.A., Blake, G.A., Tielens, A.G.G.M., Velusamy, T. & Whittet, D.C.B. (2000). Chemical Evolution of Protostellar Matter. Protostars and Planets IV, ed by V. Mannings, A. P. Boss, and S. S. Russell, (University of Arizona, Tucson), 29.

Larson, R.B. (1969). Numerical calculations of the dynamics of collapsing proto-star. MNRAS, 145, 271.

Larson, R.B. (1973a). A simple probabilistic theory of fragmentation. MNRAS, 161, 133.

Larson, R.B. (1973b). The Evolution of Protostars – Theory. Fundamentals of Cosmic Physics, 1, 1–70.

Larson, R.B. (1981). Turbulence and star formation in molecular clouds. MNRAS, 194, 809–826.

Larson, R.B. (1985). Cloud fragmentation and stellar masses. MNRAS, 214, 379–398.

Larson, R.B. (2003). The physics of star formation. Rep. Prog. Phys., 66, 1651–1697.

Larson, R.B. (2005). Thermal physics, cloud geometry and the stellar initial mass function. MNRAS, 359, 211–222.

Launay, J.M., Le Dourneuf, M. & Zeippen, C.J. (1991). The Reversible H + H$^-$ yields H$_2$(v, j) + e$^-$ Reaction - A Consistent Description of the Associative Detachment and Dissociative Attachment Processes using the Resonant Scattering Theory. A&A, 252, 842–852.

Lazarian, A. & Pogosyan, D. (2000). Velocity Modification of H I Power Spectrum. ApJ, 537, 720–748.

Le Bourlot, J., Pineau des Forêts, G. & Flower, D.R. (1999). The Cooling of Astrophysical Media by H$_2$. MNRAS, 305, 802–810.

Le Teuff, Y.H., Millar, T.J. & Markwick, A.J. (2000). The UMIST Database for Astrochemistry 1999. A&AS, 146, 157–168.

Lejeune, C. & Bastien, P. (1986). Solutions of the coagulation equation with time-dependent coagulation rates. ApJ, 309, 167–175.

Lepp, S. & Shull, J.M. (1983). The Kinetic Theory of H2 Dissociation. ApJ, 270, 578–582.

Lesieur, M. (1997). Turbulence in Fluids. (Kluwer, Dordrecht).

Li, P.S., Norman, M.L., Mac Low, M.M. & Heitsch, F. (2004). The Formation of Self-Gravitating Cores in Turbulent Magnetized Clouds. ApJ, 605, 800–818.

Li, Y., Klessen, R.S. & Mac Low, M.M. (2003). The Formation of Stellar Clusters in Turbulent Molecular Clouds: Effects of the Equation of State. ApJ, 592, 975–985.

Lin, D.N.C. & Papaloizou, J.C.B. (1996). Theory of Accretion Disks II: Application to Observed Systems. ARA&A, 34, 703–748.

Liszt, H. (2003). Gas-phase Recombination, Grain Neutralization and Cosmic-ray Ionization in Diffuse Gas. A&A, 398, 621–630.

Low, C. & Lynden-Bell, D. (1976). The minimum Jeans mass or when fragmentation must stop. MNRAS, 176, 367–390.

Lucy, L.B. (1977). A numerical approach to the testing of the fission hypothesis. AJ, 82, 1013–1024.

Ménard, F. & Duchêne, G. (2004). On the Alignment of T Tauri Stars with the Local Magnetic Field in the Taurus Molecular Cloud Complex. Ap&SS, 292, 419–425.

Mac Low, M.M. (1999). The Energy Dissipation Rate of Supersonic, Magnetohydrodynamic Turbulence in Molecular Clouds. ApJ, 524, 169–178.

Mac Low, M.M. & Klessen, R.S. (2004). Control of star formation by supersonic turbulence. Rev. Mod. Phys., 76, 125–194.

Mac Low, M.M. & Ossenkopf, V. (2000). Characterizing the structure of interstellar turbulence. A&A, 353, 339–348.

Mac Low, M.M. & Shull, J.M. (1986). Molecular Processes and Gravitational Collapse in Intergalactic Shocks. ApJ, 302, 585–589.

Mac Low, M.M., Klessen, R.S., Burkert, A. & Smith, M.D. (1998). Kinetic Energy Decay Rates of Supersonic and Super-Alfvénic Turbulence in Star-Forming Clouds. Phys. Rev. Lett., 80, 2754–2757.

Machacek, M.E., Bryan, G.L. & Abel, T. (2001). Simulations of Pregalactic Structure Formation with Radiative Feedback. ApJ, 548, 509–521.

Machacek, M.E., Bryan, G.L. & Abel, T. (2003). Effects of a soft X-ray background on structure formation at high redshift. MNRAS, 338, 273–286.

Maddox, S. (2000). The 2dF Galaxy Redshift Survey. In ASP Conf. Ser. 200: Clustering at High Redshift, 63.

Maron, J.L. & Howes, G.G. (2003). Gradient Particle Magnetohydrodynamics: A Lagrangian Particle Code for Astrophysical Magnetohydrodynamics. ApJ, 595, 564–572.

Martin, P.G., Schwarz, D.H. & Mandy, M.E. (1996). Master Equation Studies of the Collisional Excitation and Dissociation of H_2 Molecules by H Atoms. ApJ, 461, 265.

Martin, P.G., Keogh, W.J. & Mandy, M.E. (1998). Collision-induced Dissociation of Molecular Hydrogen at Low Densities. ApJ, 499, 793.

Masunaga, H. & Inutsuka, S. (2000). A Radiation Hydrodynamic Model for Protostellar Collapse. II. The Second Collapse and the Birth of a Protostar. ApJ, 531, 350–365.

Matsuda, T., Satō, H. & Takeda, H. (1969). Cooling of Pre-Galactic Gas Clouds by Hydrogen Molecule. Progress of Theoretical Physics, 42, 219–233.

McKee, C.F. & Ostriker, J.P. (1977). A theory of the interstellar medium - Three components regulated by supernova explosions in an inhomogeneous substrate. ApJ, 218, 148–169.

McKee, C.F., Zweibel, E.G., Goodman, A.A. & Heiles, C. (1993). Magnetic Fields in Star-Forming Regions - Theory. In Protostars and Planets III, ed. by E.H. Levy and J.I Lunine, (University of Arizona, Tucson), 327.

Miller, G.E. & Scalo, J.M. (1979). The initial mass function and stellar birthrate in the solar neighborhood. ApJS, 41, 513–547.

Mo, H.J. & White, S.D.M. (2002). The Abundance and Clustering of Dark Haloes in the Standard ΛCDM Cosmogony. MNRAS, 336, 112–118.

Monaghan, J.J. (1985). Particle Methods for Hydrodynamics. Comp. Phys., 3, 71–124.

Monaghan, J.J. & Lattanzio, J.C. (1985). A refined particle method for astrophysical problems. A&A, 149, 135–143.

Monaghan, J.J. & Lattanzio, J.C. (1991). A simulation of the collapse and fragmentation of cooling molecular clouds. ApJ, 375, 177–189.

Moore, B., Ghigna, S., Governato, F., Lake, G., Quinn, T., Stadel, J. & Tozzi, P. (1999). Dark Matter Substructure within Galactic Halos. ApJL, 524, L19–L22.

Moseley, J., Aberth, W. & Peterson, J.R. (1970). $H^+ + H^-$ Mutual Neutralization Cross Section Obtained with Superimposed Beams. Phys. Rev. Lett., 24, 435.

Motte, F., André, P. & Neri, R. (1998). The initial conditions of star formation in the rho Ophiuchi main cloud: wide-field millimeter continuum mapping. A&A, 336, 150–172.

Mouschovias, T.C. & Paleologou, E.V. (1979). The angular momentum problem and magnetic braking - an exact time-dependent solution. ApJ, 230, 204–222.

Mouschovias, T.C. & Paleologou, E.V. (1980). Magnetic braking of an aligned rotator during star formation - an exact, time-dependent solution. ApJ, 237, 877–899.

Murray, S.D. & Lin, D.N.C. (1996). Coalescence, Star Formation, and the Cluster Initial Mass Function. ApJ, 467, 728.

Myers, P.C. (1978). A compilation of interstellar gas properties. ApJ, 225, 380–389.

Myers, P.C. (1999). Physical Conditions in Nearby Molecular Clouds. In NATO ASIC Proc. 540: The Origin of Stars and Planetary Systems, 67.

Myers, P.C., Fuller, G.A., Goodman, A.A. & Benson, P.J. (1991). Dense cores in dark clouds. VI - Shapes. ApJ, 376, 561–572.

Myhill, E.A. & Kaula, W.M. (1992). Numerical models for the collapse and fragmentation of centrally condensed molecular cloud cores. ApJ, 386, 578–586.

Nahar, S.N. (1999). Electron-Ion Recombination Rate Coefficients, Photoionization Cross Sections, and Ionization Fractions for Astrophysically Abundant Elements. II. Oxygen Ions. ApJS, 120, 131–145.

Nahar, S.N. (2000). Electron-Ion Recombination Rate Coefficients and Photoionization Cross Sections for Astrophysically Abundant Elements. III. Si-Sequence Ions: Si I, S III, Ar V, Ca VII, and Fe XIII. ApJS, 126, 537–550.

Nahar, S.N. & Pradhan, A.K. (1997). Electron-Ion Recombination Rate Coefficients, Photoionization Cross Sections, and Ionization Fractions for Astrophysically Abundant Elements. I. Carbon and Nitrogen. ApJS, 111, 339.

Nakano, T., Hasegawa, T. & Norman, C. (1995). The Mass of the Star Formed in a Cloud Core. Ap&SS, 224, 523–524.

Navarro, J.F. & Steinmetz, M. (2000). The Core Density of Dark Matter Halos: A Critical Challenge to the ΛCDM Paradigm? ApJ, 528, 607–611.

Navarro, J.F., Frenk, C.S. & White, S.D.M. (1996). The Structure of Cold Dark Matter Halos. ApJ, 462, 563.

Navarro, J.F., Frenk, C.S. & White, S.D.M. (1997). A Universal Density Profile from Hierarchical Clustering. ApJ, 490, 493.

Nisini, B., Massi, F., Vitali, F., Giannini, T., Lorenzetti, D., Di Paola, A., Codella, C., D'Alessio, F. & Speziali, R. (2001). Multiple H_2 protostellar jets in the bright-rimmed globule IC 1396-N. A&A, 376, 553–560.

Norman, C.A. & Ferrara, A. (1996). The Turbulent Interstellar Medium: Generalizing to a Scale-dependent Phase Continuum. ApJ, 467, 280.

Nozawa, T., Kozasa, T., Umeda, H., Maeda, K. & Nomoto, K. (2003). Dust in the Early Universe: Dust Formation in the Ejecta of Population III Supernovae. ApJ, 598, 785–803.

Oh, S.P. & Haiman, Z. (2003). Fossil H II Regions: Self-limiting Star Formation at High Redshift. MNRAS, 346, 456–472.

O'Shea, B.W., Abel, T., Whalen, D. & Norman, M.L. (2005). Forming a Primordial Star in a Relic H II Region. ApJL, 628, L5–L8.

Ossenkopf, V. & Mac Low, M.M. (2002). Turbulent velocity structure in molecular clouds. A&A, 390, 307–326.

Ossenkopf, V., Klessen, R.S. & Heitsch, F. (2001). On the structure of self-gravitating molecular clouds. A&A, 379, 1005–1016.

Ottino, J.M. (1989). The Kinematics of Mixing. (Cambridge University Press, Cambridge, UK).

Padoan, P. (1995). Supersonic turbulent flows and the fragmentation of a cold medium. MNRAS, 277, 377–388.

Padoan, P. & Nordlund, Å. (2002). The Stellar Initial Mass Function from Turbulent Fragmentation. ApJ, 576, 870–879.

Padoan, P., Nordlund, A. & Jones, B.J.T. (1997). The universality of the stellar initial mass function. MNRAS, 288, 145–152.

Palla, F. & Zinnecker, H. (1987). Non-equilibrium Cooling of a Hot Primordial Gas Cloud. In Starbursts and Galaxy Evolution, Proceedings of the Twenty-second Moriond Astrophysics Meeting, (Editions Frontieres, Gif-sur-Yvette, France), 533–540.

Panis, J. & Pérault, M. (1998). Numerical study of the effective viscosity and pressure in perturbed turbulent flows. Physics of Fluids, 10, 3111–3125.

Papaloizou, J.C.B. & Lin, D.N.C. (1995). Theory Of Accretion Disks I: Angular Momentum Transport Processes. ARA&A, 33, 505–540.

Peebles, P.J.E. (1980). The Large-scale Structure of the Universe. Research supported by the National Science Foundation, (Princeton University Press, Princeton, NJ).

Peebles, P.J.E. (1993). Principles of Physical Cosmology. Princeton Series in Physics, (Princeton University Press, Princeton, NJ).

Peebles, P.J.E. & Dicke, R.H. (1968). Origin of the Globular Star Clusters. ApJ, 154, 891.

Penston, M.V. (1969a). Dynamics of self-gravitating gaseous spheres-III. Analytical results in the free-fall of isothermal cases. MNRAS, 144, 425.

Penston, M.V. (1969b). Dynamics of self-gravitating gaseous spheres-III. Analytical results in the free-fall of isothermal cases. MNRAS, 144, 425.

Pettini, M. (1999). Element Abundances at High Redshifts. In Chemical Evolution from Zero to High Redshift, ed. by J.R. Walsh and M.R. Rosa, (Springer, Berlin), 233.

Pirogov, L., Zinchenko, I., Caselli, P., Johansson, L.E.B. & Myers, P.C. (2003). N_2H^+(1-0) survey of massive molecular cloud cores. A&A, 405, 639–654.

Porter, D.H. & Woodward, P.R. (1994). High-resolution simulations of compressible convection using the piecewise-parabolic method. ApJS, 93, 309–349.

Porter, D.H., Pouquet, A. & Woodward, P.R. (1992). Three-dimensional supersonic homogeneous turbulence - A numerical study. Physical Review Letters, 68, 3156–3159.

Porter, D.H., Pouquet, A. & Woodward, P.R. (1994). Kolmogorov-like spectra in decaying three-dimensional supersonic flows. Physics of Fluids, 6, 2133–2142.

Poulaert, G., Brouillard, F., Claeys, W., McGowan, J.W. & Van Wassenhove, G. (1978). H_2^+ Formation in Low Energy $H^+ - H^-$ Collisions . Journal of Physics B Atomic Molecular Physics, 11, L671–L673.

Press, W.H. & Schechter, P. (1974). Formation of Galaxies and Clusters of Galaxies by Self-Similar Gravitational Condensation. ApJ, 187, 425–438.

Price, D.J. & Monaghan, J.J. (2004). Smoothed Particle Magnetohydrodynamics: Some Shocking Results. Ap&SS, 292, 279–283.

Price, N.M. & Podsiadlowski, P. (1995). Dynamical interactions between young stellar objects and a collisional model for the origin of the stellar mass spectrum. MNRAS, 273, 1041–1068.

Ramaker, D.E. & Peek, J.M. (1976). Molecule Formation in Tenuous Media: Quantum Effects in Spontaneous Radiative Association. Phys. Rev. A, 13, 58–64.

Reipurth, B. & Clarke, C. (2001). The Formation of Brown Dwarfs as Ejected Stellar Embryos. AJ, 122, 432–439.

Ricotti, M. & Ostriker, J.P. (2004). Reionization, chemical enrichment and seed black holes from the first stars: is Population III important? MNRAS, 350, 539–551.

Ricotti, M., Gnedin, N.Y. & Shull, J.M. (2002). The Fate of the First Galaxies. I. Self-consistent Cosmological Simulations with Radiative Transfer. ApJ, 575, 33–48.

Saito, M., Kawabe, R., Ishiguro, M., Miyama, S.M., Hayashi, M., Handa, T., Kitamura, Y. & Omodaka, T. (1995). Aperture Synthesis 12CO and 13CO Observations of DM Tauri: 350 AU Radius Circumstellar Gas Disk. ApJ, 453, 384.

Salpeter, E.E. (1955). The Luminosity Function and Stellar Evolution. ApJ, 121, 161.

Sasao, T. (1973). On the Generation of Density Fluctuations Due to Turbulence in Self-Gravitating Media. PASJ, 25, 1.

Saslaw, W.C. & Zipoy, D. (1967). Molecular Hydrogen in Pre-galactic Gas Clouds. Nature, 216, 976.

Savin, D.W., Krstić, P.S., Haiman, Z. & Stancil, P.C. (2004). Rate Coefficient for $H^+ + H_2(X^1\Sigma_g^+,$ $\nu = 0, J = 0)$ -> $H(1s) + H_2^+$ Charge Transfer and Some Cosmological Implications. ApJL, 606, L167; erratum ApJ, 607, L147.

Scalo, J. (1998). The IMF Revisited: A Case for Variations. In ASP Conf. Ser. 142: The Stellar Initial Mass Function (38th Herstmonceux Conference), ed. G. Gilmore & D. Howell (Astron. Soc. Pac., San Francisco), 201.

Scalo, J. & Elmegreen, B.G. (2004). Interstellar Turbulence II: Implications and Effects. ARA&A, 42, 275–316.

Scalo, J., Vazquez-Semadeni, E., Chappell, D. & Passot, T. (1998). On the Probability Density Function of Galactic Gas. I. Numerical Simulations and the Significance of the Polytropic Index. ApJ, 504, 835.

Scalo, J.M. (1986). The stellar initial mass function. Fundamentals of Cosmic Physics, 11, 1–278.

Schmeja, S. & Klessen, R.S. (2004). Protostellar mass accretion rates from gravoturbulent fragmentation. A&A, 419, 405–417.

Schneider, I.F., Dulieu, O., Giusti-Suzor, A. & Roueff, E. (1994). Dissociate Recombination of H_2^+ Molecular Ions in Hydrogen Plasmas between 20 K and 4000 K. ApJ, 424, 983; erratum ApJ, 486, 580.

Schneider, R., Ferrara, A. & Salvaterra, R. (2004). Dust formation in very massive primordial supernovae. MNRAS, 351, 1379–1386.

Schneider, S. & Elmegreen, B.G. (1979). A Catalog of Dark Globular Filaments. ApJS, 41, 87–95.

Seljak, U. (2000). Analytic Model for Galaxy and Dark Matter Clustering. MNRAS, 318, 203–213.

Seljak, U. & Zaldarriaga, M. (1996). A Line-of-Sight Integration Approach to Cosmic Microwave Background Anisotropies. ApJ, 469, 437.

Shang, H., Li, Z.Y., Shu, F.H. & Hirano, N. (2006). Jets and Bipolar Outflows from Young Stars - Theory and Observational Tests. In Protostars and Planets V, in preparation.

Shapiro, P.R. & Kang, H. (1987). Hydrogen Molecules and the Radiative Cooling of Pregalactic Shocks. ApJ, 318, 32–65.

She, Z. & Leveque, E. (1994). Universal scaling laws in fully developed turbulence. Physical Review Letters, 72, 336–339.

Shu, F.H. (1977). Self-similar collapse of isothermal spheres and star formation. ApJ, 214, 488–497.

Shu, F.H., Adams, F.C. & Lizano, S. (1987). Star formation in molecular clouds - Observation and theory. ARA&A, 25, 23–81.

Shu, F.H., Allen, A., Shang, H., Ostriker, E.C. & Li, Z.Y. (1999). Low-Mass Star Formation: Theory. In NATO ASIC Proc. 540: The Origin of Stars and Planetary Systems, 193.

Shu, F.H., Li, Z.Y. & Allen, A. (2004). Does Magnetic Levitation or Suspension Define the Masses of Forming Stars? ApJ, 601, 930–951.

Silk, J. (1995). A theory for the initial mass function. ApJL, 438, L41–L44.

Silk, J. & Takahashi, T. (1979). A statistical model for the initial stellar mass function. ApJ, 229, 242–256.

Simon, M. (1992). Multiplicity Among the Young Stars. In ASP Conf. Ser. 32: IAU Colloq. 135: Complementary Approaches to Double and Multiple Star Research, 41.

Spaans, M. & Silk, J. (2000). The Polytropic Equation of State of Interstellar Gas Clouds. ApJ, 538, 115–120.

Spaans, M. & Silk, J. (2005). The Polytropic Equation of State of Primordial Gas Clouds. ApJ, 626, 644–648.

Spergel et al. (2003). First-Year Wilkinson Microwave Anisotropy Probe (WMAP) Observations: Determination of Cosmological Parameters. ApJS, 148, 175–194.

Spitzer, L. (1968). Diffuse matter in space. (Interscience Publication, New York).

Springel, V., Yoshida, N. & White, S.D.M. (2001). GADGET: a code for collisionless and gasdynamical cosmological simulations. New Astronomy, 6, 79–117.

Stahler, S.W. & Palla, F. (2004). The Formation of Stars . (Wiley-VCH, Weinheim).

Stancil, P.C., Havener, C.C., Krstic, P.S., Schultz, D.R., Kimura, M., Gu, J.P., Hirsch, G., Buenker, R.J. & Zygelman, B. (1998). Charge Transfer in Collisions of C^+ with H and H^+ with C. ApJ, 502, 1006.

Stancil, P.C., Schultz, D.R., Kimura, M., Gu, J.P., Hirsch, G. & Buenker, R.J. (1999). Charge Transfer in Collisions of O^+ with H and H^+ with O. A&AS, 140, 225–234.

Stibbe, D.T. & Tennyson, J. (1999). Rates for the Electron Impact Dissociation of Molecular Hydrogen. ApJL, 513, L147–L150.

Stolte, A., Brandner, W., Grebel, E.K., Lenzen, R. & Lagrange, A.M. (2005). The Arches Cluster: Evidence for a Truncated Mass Function? ApJL, 628, L113–L117.

Stone, J.M., Ostriker, E.C. & Gammie, C.F. (1998). Dissipation in Compressible Magnetohydrodynamic Turbulence. ApJL, 508, L99–L102.

Stutzki, J. & Guesten, R. (1990). High spatial resolution isotopic CO and CS observations of M17 SW - The clumpy structure of the molecular cloud core. ApJ, 356, 513–533.

Tafalla, M., Myers, P.C., Caselli, P. & Walmsley, C.M. (2004). On the internal structure of starless cores. I. Physical conditions and the distribution of CO, CS, N_2H^+, and NH_3 in L1498 and L1517B. A&A, 416, 191–212.

Tarafdar, S.P., Prasad, S.S., Huntress, W.T., Villere, K.R. & Black, D.C. (1985). Chemistry in Dynamically Evolving Clouds. ApJ, 289, 220–237.

Tegmark, M., Silk, J., Rees, M.J., Blanchard, A., Abel, T. & Palla, F. (1997). How Small Were the First Cosmological Objects? ApJ, 474, 1.

Tielens, A.G.G.M. (1991). Characterstics of Interstellar and Circumstellar Dust (invited Review). In ASSL Vol. 173: IAU Colloq. 126: Origin and Evolution of Interplanetary Dust, ed. by A. C. Levasseur-Regourd and H. Hasegawa. (Kluwer, Dordrecht), 405.

Todini, P. & Ferrara, A. (2001). Dust Formation in Primordial Type II Supernovae. MNRAS, 325, 726–736.

Tomisaka, K. (1996). Collapse and Fragmentation of Cylindrical Magnetized Clouds: Simulation with Nested Grid Scheme. PASJ, 48, 701–717.

Troland, T.H., Crutcher, R.M., Goodman, A.A., Heiles, C., Kazes, I. & Myers, P.C. (1996). The Magnetic Fields in the Ophiuchus and Taurus Molecular Clouds. ApJ, 471, 302.

Truelove, J.K., Klein, R.I., McKee, C.F., Holliman, J.H., Howell, L.H. & Greenough, J.A. (1997). The Jeans Condition: A New Constraint on Spatial Resolution in Simulations of Isothermal Self-gravitating Hydrodynamics. ApJL, 489, L179.

Turner, J.A., Chapman, S.J., Bhattal, A.S., Disney, M.J., Pongracic, H. & Whitworth, A.P. (1995). Binary star formation: gravitational fragmentation followed by capture. MNRAS, 277, 705–726.

Vázquez-Semadeni, E., Ostriker, E.C., Passot, T., Gammie, C.F. & Stone, J.M. (2000). Compressible MHD Turbulence: Implications for Molecular Cloud and Star Formation. Protostars and Planets IV, ed. V. Mannings, A. P. Boss & S. S. Russell (Univ. of Arizona Press, Tucson), 3.

Vázquez-Semadeni, E., Kim, J., Shadmehri, M. & Ballesteros-Paredes, J. (2005). The Lifetimes and Evolution of Molecular Cloud Cores. ApJ, 618, 344–359.

van Dishoeck, E.F. (1988). Photodissociation and Photoionization Processes, 49–72. Rate Coefficients in Astrochemistry (A90-18793 06-90), eds Millar, T. J., Williams, D. A.,(Kluwer Academic Publishers, Dordrecht and Norwell, MA).

van Dishoek, E.F. & Hogerheijde, M.R. (2000). Models and Observations of the Chemistry Near Young Stellar Objects. In The Origin of Stars and Planetary Systems, ed. by C.J. Lada and N.D. Kylafis,(Kluwer, Dordrecht), 163–241.

van Dishoek, E.F., Blake, G.A., Draine, B.T. & Lunine, J.I. (1993). The Chemical Evolution of Protostellar and Protoplanetary Matter. In Protostars and Planets III, ed. by E.H. Levy and J.I Lunine, (University of Arizona, Tucson), 163–241.

Verner, D.A., Ferland, G.J., Korista, K.T. & Yakovlev, D.G. (1996). Atomic Data for Astrophysics. II. New Analytic FITS for Photoionization Cross Sections of Atoms and Ions. ApJ, 465, 487.

Verschuur, G.L. (1995). Zeeman Effect Observations of H i Emission Profiles. I. Magnetic Field Limits for Three Regions Based on Observations Corrected for Polarized Beam Structure. ApJ, 451, 624.

von Weizsäcker, C.F. (1943). Über die Entstehung des Planetensystems. Z. f. Astrophys., 22, 319.

von Weizsäcker, C.F. (1951). The Evolution of Galaxies and Stars. ApJ, 114, 165.

Voronov, G.S. (1997). A Practical Fit Formula for Ionization Rate Coefficients of Atoms and Ions by Electron Impact: Z = 1-28. Atomic Data and Nuclear Data Tables, 65, 1.

Wada, K., Spaans, M. & Kim, S. (2000). Formation of Cavities, Filaments, and Clumps by the Nonlinear Development of Thermal and Gravitational Instabilities in the Interstellar Medium under Stellar Feedback. ApJ, 540, 797–807.

Ward-Thompson, D., Scott, P.F., Hills, R.E. & Andre, P. (1994). A Submillimetre Continuum Survey of Pre Protostellar Cores. MNRAS, 268, 276.

Weigert, A. & Wendker, H.J. (1989). Astronomie und Astrophysik - ein Grundkurs. (VCH Verlagsgesellschaft, Weinheim).

Weingartner, J.C. & Draine, B.T. (2001). Electron-Ion Recombination on Grains and Polycyclic Aromatic Hydrocarbons. ApJ, 563, 842–852.

Whitworth, A. & Summers, D. (1985). Self-similar condensation of spherically symmetric self-gravitating isothermal gas clouds. MNRAS, 214, 1–25.

Whitworth, A.P., Chapman, S.J., Bhattal, A.S., Disney, M.J., Pongracic, H. & Turner, J.A. (1995). Binary star formation: accretion-induced rotational fragmentation. MNRAS, 277, 727–746.

Whitworth, A.P., Bhattal, A.S., Francis, N. & Watkins, S.J. (1996). Star formation and the singular isothermal sphere. MNRAS, 283, 1061–1070.

Williams, J.P., Blitz, L. & McKee, C.F. (2000). The Structure and Evolution of Molecular Clouds: from Clumps to Cores to the IMF. Protostars and Planets IV, ed. V. Mannings, A. P. Boss & S. S. Russell (Univ. of Arizona Press, Tucson), 97.

Wirtz, C. (1924). De Sitters Kosmologie und die Radialbewegungen der Spiralnebel. Astronomische Nachrichten, 222, 21.

Wishart, A.W. (1979). The Bound-free Photo-detachment Cross-section of H^-. MNRAS, 187, 59P–610P.

Wolfire, M.G., Hollenbach, D., McKee, C.F., Tielens, A.G.G.M. & Bakes, E.L.O. (1995). The neutral atomic phases of the interstellar medium. ApJ, 443, 152–168.

Wu, K., Lahav, O. & Rees, M. (1999). The large-scale Smoothness of the Universe. Nature, 397, 225.

Wuchterl, G. & Klessen, R.S. (2001). The First Million Years of the Sun: A Calculation of the Formation and Early Evolution of a Solar Mass Star. ApJL, 560, L185–L188.

York et al. (2000). The Sloan Digital Sky Survey: Technical Summary. AJ, 120, 1579–1587.

Yoshida, N., Abel, T., Hernquist, L. & Sugiyama, N. (2003). Simulations of Early Structure Formation: Primordial Gas Clouds. ApJ, 592, 645–663.

Ziegler, U. (2005). Self-gravitational Adaptive Mesh Magnetohydrodynamics with the NIRVANA Code. A&A, 435, 385–395.

Zinnecker, H. (1984). Star formation from hierarchical cloud fragmentation - A statistical theory of the log-normal Initial Mass Function. MNRAS, 210, 43–56.

Zinnecker, H. (1990). Physical Processes in Fragmentation and Star Formation, ed. R. Capuzzo-Dolcetta and C. Chiosi (Kluwer, Dordrecht), 201.

Zinnecker, H. (2004). The stellar mass and angular momentum problem in star formation. In Revista Mexicana de Astronomia y Astrofisica Conference Series, 77–80.

Zucconi, A., Walmsley, C.M. & Galli, D. (2001). The dust temperature distribution in prestellar cores. A&A, 376, 650–662.

Index

A
angular momentum 21

C
CDM 25
cold dark matter 25
cooling time 39
cosmic rays..........................17, 53
cosmology 23

D
dark matter halos 26

E
equation of state
    isothermal 35
    polytropic...........................35

F
free-fall time 38

G
GADGET 40
gravoturbulence..........................29

H
Hubble time 39
Hubble's law 24

I
IMF 17
initial mass function 17
interstellar medium 13
ISM 13
isothermal equation of state..............35

J
Jeans mass...............................31

N
numerical scheme 40

P
photoelectric heating....................17
polytropic equation of state 35
population II............................28
population III.............28, 53, 96, 112 ff
Press-Schechter Mass Function 26

R
redshift 24
reionization epoch...........28, 92, 94, 96 ff

S
self-shielding............................53
sink particles............................43
smoothed particle hydrodynamics 40
sound speed....................31, 35, 37 ff
sound-crossing time 38
sph 40

T
thermal properties 15
turbulence 31
turbulent crossing time..................39
turbulent driving........................49

V
velocity dispersion 58
virial radius.............................101
virial temperature 111

viscosity 32

W

WMAP 25

Z

Zeeman splitting 14

Die VDM Verlagsservicegesellschaft sucht für wissenschaftliche Verlage abgeschlossene und herausragende

Dissertationen, Habilitationen, Diplomarbeiten, Master Theses, Magisterarbeiten usw.

für die kostenlose Publikation als Fachbuch.

Sie verfügen über eine Arbeit, die hohen inhaltlichen und formalen Ansprüchen genügt, und haben Interesse an einer honorarvergüteten Publikation?

Dann senden Sie bitte erste Informationen über sich und Ihre Arbeit per Email an *info@vdm-vsg.de*.

Sie erhalten kurzfristig unser Feedback!

VDM Verlagsservicegesellschaft mbH
Dudweiler Landstr. 99   Telefon +49 681 3720 174
D - 66123 Saarbrücken   Fax   +49 681 3720 1749
www.vdm-vsg.de

Die VDM Verlagsservicegesellschaft mbH vertritt

Printed by Books on Demand GmbH, Norderstedt / Germany